Florian Meirer

Applications of SR-TXRF Analysis in XAS

Florian Meirer

Applications of SR-TXRF Analysis in XAS

Applications of Synchrotron Radiation induced Total Reflection X-Ray Fluorescence Analysis in Absorption Spectroscopy

Südwestdeutscher Verlag für Hochschulschriften

Impressum/Imprint (nur für Deutschland/ only for Germany)
Bibliografische Information der Deutschen Nationalbibliothek: Die Deutsche Nationalbibliothek
verzeichnet diese Publikation in der Deutschen Nationalbibliografie; detaillierte bibliografische
Daten sind im Internet über http://dnb.d-nb.de abrufbar.
Alle in diesem Buch genannten Marken und Produktnamen unterliegen warenzeichen-, marken-
oder patentrechtlichem Schutz bzw. sind Warenzeichen oder eingetragene Warenzeichen der
jeweiligen Inhaber. Die Wiedergabe von Marken, Produktnamen, Gebrauchsnamen,
Handelsnamen, Warenbezeichnungen u.s.w. in diesem Werk berechtigt auch ohne besondere
Kennzeichnung nicht zu der Annahme, dass solche Namen im Sinne der Warenzeichen- und
Markenschutzgesetzgebung als frei zu betrachten wären und daher von jedermann benutzt
werden dürften.

Verlag: Südwestdeutscher Verlag für Hochschulschriften Aktiengesellschaft & Co. KG
Dudweiler Landstr. 99, 66123 Saarbrücken, Deutschland
Telefon +49 681 37 20 271-1, Telefax +49 681 37 20 271-0, Email: info@svh-verlag.de
Zugl.: Vienna, Vienna University of Technology, Diss., 2008

Herstellung in Deutschland:
Schaltungsdienst Lange o.H.G., Zehrensdorfer Str. 11, D-12277 Berlin
Books on Demand GmbH, Gutenbergring 53, D-22848 Norderstedt
Reha GmbH, Dudweiler Landstr. 99, D- 66123 Saarbrücken
ISBN: 978-3-8381-0792-9

Imprint (only for USA, GB)
Bibliographic information published by the Deutsche Nationalbibliothek: The Deutsche
Nationalbibliothek lists this publication in the Deutsche Nationalbibliografie; detailed
bibliographic data are available in the Internet at http://dnb.d-nb.de.
Any brand names and product names mentioned in this book are subject to trademark, brand or
patent protection and are trademarks or registered trademarks of their respective holders. The
use of brand names, product names, common names, trade names, product descriptions etc.
even without
a particular marking in this works is in no way to be construed to mean that such names may be
regarded as unrestricted in respect of trademark and brand protection legislation and could thus
be used by anyone.

Publisher:
Südwestdeutscher Verlag für Hochschulschriften Aktiengesellschaft & Co. KG
Dudweiler Landstr. 99, 66123 Saarbrücken, Germany
Phone +49 681 37 20 271-1, Fax +49 681 37 20 271-0, Email: info@svh-verlag.de

Copyright © 2008 Südwestdeutscher Verlag für Hochschulschriften Aktiengesellschaft & Co. KG
and licensors
All rights reserved. Saarbrücken 2008

Produced in USA and UK by:
Lightning Source Inc., 1246 Heil Quaker Blvd., La Vergne, TN 37086, USA
Lightning Source UK Ltd., Chapter House, Pitfield, Kiln Farm, Milton Keynes, MK11 3LW, GB
BookSurge, 7290 B. Investment Drive, North Charleston, SC 29418, USA
ISBN: 978-3-8381-0792-9

für Eleonore

For in and out, above, about, below,
'Tis nothing but a Magic Shadow-show
  Play'd in a Box whose Candle is the Sun,
Round which we Phantom Figures come and go.
  Rubáiyat of Omar Khayyám, Quatrain XLVI

Abstract

Synchrotron radiation induced TXRF (SR-TXRF) is a microanalytical technique which offers detection limits in the fg range for most elements. The technique can be coupled to X-ray Absorption Spectroscopy (XAS) to gain information on the chemical environment of the specific elements of interest at an ultra trace level. The combination of these techniques has been applied to various analytical problems.

X-Ray Absorption Near Edge Structure (XANES) analysis in total reflection geometry was used to determine the chemical state of Fe contaminations on a silicon wafer surface. The ability to characterise chemically the contamination on silicon wafers is of critical importance to the semiconductor industry. It provides information on possible unwanted chemical processes taking place on the wafer surface and helps in determining the true source of the contamination problem. This type of information is not readily accessible with standard laboratory equipment. Main purpose of the study was to test the method for a contamination issue as it could appear in a microelectronic VLSI (Very-Large-Scale Integration) production fab.

To understand the effects of aerosols on human health and global climate a detailed understanding of sources, transport, and fate as well as of the physical and chemical properties of atmospheric particles is necessary. An analysis of aerosols should therefore provide information about size and elemental composition of the particles and - if desired - deliver information about the chemical state of a specific element of interest in the particles. Using the combination of SR-TXRF and XANES analysis it was possible to investigate the elemental composition of size fractioned atmospheric aerosols and the oxidation state of Fe in the aerosols even for small sample amounts due to short aerosol collection times.

The applicability of this combined technique was further tested for the determination of the arsenic species in cucumber (Cucumis sativus L.) xylem saps. The speciation of arsenic is relevant because the toxicity of arsenic differs considerably dependent on the oxidation state and chemical form and it is known that plants have the capability to change the oxidation state of arsenic.

During this work a damping of the oscillations of the absorption fine structure was observed when measuring samples with higher concentrations. It was assumed that the reason is a self absorption effect (absorption along the path of the incident beam) which occurs due to the extreme total reflection geometry. The influence of self absorption effects on TXRF-XANES

Abstract

measurements was investigated by comparing measurements with theoretical calculations. Additionally the inverse TXRF geometry – the grazing exit setup – was tested for its applicability to XANES analysis and applied to gain a better understanding of the above mentioned self absorption effect.

In the framework of this thesis it could be shown that TXRF analysis in combination with XANES analysis is a powerful and multifunctional method to perform chemical speciation studies at trace element levels.

Part of the work reported in this thesis appears in the following publications and HASYLAB annual reports:

[1] F. Meirer, G. Pepponi, C. Streli, P. Wobrauschek, V.G. Mihucz, G. Záray, V. Czech, J.A.C. Broekaert, U.E.A. Fittschen, G. Falkenberg, Application of synchrotron-radiation-induced TXRF-XANES for arsenic speciation in cucumber (*Cucumis sativus L.*) xylem sap, X-Ray Spectrometry 36 (2007) 408-412.

[2] F. Meirer, G. Pepponi, C. Streli, P. Wobrauschek, P. Kregsamer, N. Zöger, G. Falkenberg, Parameter study of self-absorption effects in TXRF-XANES analysis of arsenic, Spectrochimica Acta Part B, accepted (2008).

[3] F. Meirer, C. Streli, G. Pepponi, P. Wobrauschek, M.A. Zaitz, C. Horntrich, G. Falkenberg, Feasibility study of SR-TXRF-XANES analysis for iron contaminations on a silicon wafer surface, submitted to Surface and Interface Analysis (2008).

[4] V. Groma, J. Osan, A. Alsecz, S. Török, F. Meirer, C. Streli, P. Wobrauschek, G. Falkenberg, Trace element analysis of airport related aerosols using SR-TXRF, IDŐJÁRÁS - Journal of the Hungarian Meteorological Service 112, accepted (2008).

[5] U.E.A. Fittschen, F. Meirer, C. Streli, P. Wobrauschek, J. Thiele, G. Falkenberg, G. Pepponi, Characterization of Atmospheric Aerosols using SR-TXRF and Fe K-edge TXRF-XANES, submitted to Spectrochimica Acta Part B (2008).

[6] C. Streli, P. Wobrauschek, F. Meirer, G. Pepponi, Synchrotron radiation induced TXRF- a critical review, Journal of Analytical Atomic Spectrometry DOI: 10.1039/b719508g (2008).

[7] F. Meirer, G. Pepponi, C. Streli, P. Wobrauschek, V.G. Mihucz, G. Zaray, V. Czech, J.A.C. Broekaert, U.E.A. Fittschen, G. Falkenberg, Arsenic speciation in cucumber (Cucumis sativus L.) xylem sap by K-edge TXRF-XANES, HASYLAB Annual Report (2006).

[8] F. Meirer, C. Streli, P. Wobrauschek, N. Zoeger, C. Jokubonis, G. Pepponi, G. Falkenberg, J. Osan, S. Török, V. Groma, U.E.A. Fittschen, J.A.C. Broekaert, V.G. Mihucz, G. Zaray, V. Czech, J. Hofstätter, P. Roschger, R. Simon, Recent results of the ATI X-Ray group with SR-XRF, XRF Newsletter of the IAEA Laboratories 12 (2006) 7-13.

[9] F. Meirer, G. Pepponi, C. Streli, P. Wobrauschek, C. Horntrich, M.A. Zaitz, G. Falkenberg, Characterization of iron-contaminations on silicon wafer surface, HASYLAB Annual Report (2007).

[10] F. Meirer, G. Pepponi, C. Streli, P. Wobrauschek, P. Kregsamer, N. Zoeger, G. Falkenberg, Parameter study of self-absorption effects in TXRF-XANES analysis of arsenic, HASYLAB Annual Report (2007).

[11] F. Meirer, G. Pepponi, C. Streli, P. Wobrauschek, N. Zoeger, K. Rickers, A new Grazing-Exit-XRF setup at HASYLAB beamline L, HASYLAB Annual Report (2007).

[12] J. Osan, S. Török, V. Groma, C. Streli, P. Wobrauschek, F. Meirer, G. Falkenberg, Trace element analysis of fine aerosol particles with high time resolution using SR-TXRF, HASYLAB Annual Report (2005).

[13] C. Streli, P. Wobrauschek, C. Jokubonis, F. Meirer, J.A.C. Broekaert, U.E.A. Fittschen, G. Pepponi, G. Zaray, G. Falkenberg, SR-TXRF at Beamline L: Performance and XANES Applications, HASYLAB Annual Report (2005).

[14] G. Pepponi, C. Streli, P. Wobrauschek, F. Meirer, G. Zaray, V.G. Mihucz, J.A.C. Broekaert, U.E.A. Fittschen, G. Falkenberg, SR-TXRF at Beamline L: XANES on As in Xylem Sap of Cucumber Plants, HASYLAB Annual Report (2005).

[15] V. Groma, F. Meirer, J. Osan, S. Török, C. Streli, P. Wobrauschek, G. Falkenberg, Trace element analysis of urban and background aerosols using SR-TXRF, HASYLAB Annual Report (2006).

[16] U.E.A. Fittschen, F. Meirer, C. Streli, P. Wobrauschek, G. Falkenberg, G. Pepponi, J.A.C. Broekaert, J. Thiele, Characterization of Atmospheric Aerosols using SR-TXRF and Fe K-edge TXRF-XANES, HASYLAB Annual Report (2007).

The most recent work on the application of the grazing exit setup will be published soon:
- F. Meirer, G. Pepponi, C. Streli, P. Wobrauschek, N. Zoeger, K. Rickers, Characterization of Arsenic droplet samples with GE-XRF and GE-XANES, to be published (2008)

Abstract

Poster (2 best poster awards) and talks related to parts of this work have been presented at the following conferences:

- EXRS 2006, Paris, France; 2 Poster:
    - F. Meirer, G. Pepponi, C. Streli, P. Wobrauschek, V.G. Mihucz, G. Záray, V. Czech, J.A.C. Broekaert, U.E.A. Fittschen, G. Falkenberg, Arsenic speciation in cucumber (*Cucumis sativus L.*) xylem sap by K-Edge TXRF-XANES
    - V. Groma, J. Osan, A. Alsecz, S. Török, C. Streli, P. Wobrauschek, F. Meirer, G. Falkenberg, Trace elemental analysis of fine aerosol particles with high time resolution using SR-TXRF
- Denver X-Ray Conference 06, Denver, USA; Poster (Best Poster Award) and Talk (U.E.A. Fittschen)
    - G. Pepponi, F. Meirer, C. Streli, P. Wobrauschek, V.G. Mihucz, G. Záray, V. Czech, J.A.C. Broekaert, U.E.A. Fittschen, G. Falkenberg, Arsenic speciation in cucumber (*Cucumis sativus L.*) xylem sap by K-Edge TXRF-XANES
    - U.E.A. Fittschen, J.A.C. Broekaert, S. Hauschild, D. Rehder, S. Förster, C. Streli, P. Wobrauschek, C. Jokubonis, F. Meirer, G. Pepponi, G. Lammel, G. Falkenberg, Analysis of atmospheric aerosols and cell cultures with SR-TXRF: new direct calibration using pico-droplets (pL) generated by ink-jet printers and speciation of Iron with SR-TXRF-XANES
- Conference of the Austrian Physical Society 2006, Graz, Austria; Poster
    - F. Meirer, G. Pepponi, C. Streli, P. Wobrauschek, V.G. Mihucz, G. Záray, V. Czech, J.A.C. Broekaert, U.E.A. Fittschen, G. Falkenberg, Arsenic speciation in cucumber (*Cucumis sativus L.*) xylem sap by K-Edge TXRF-XANES
- TXRF Conference 2007, Trento, Italy, Talk (Meirer) and 2 Poster
    - F. Meirer, G. Pepponi, C. Streli, P. Wobrauschek, C. Horntrich, J. Broekaert, U. E. A. Fittschen, G. Falkenberg, Parameter study of self-absorption effects in TXRF-XANES measurements
    - F. Meirer, G. Pepponi, C. Streli, P. Wobrauschek, V.G. Mihucz, G. Záray, V. Czech, J.A.C. Broekaert, U.E.A. Fittschen, G. Falkenberg, Arsenic speciation in cucumber (*Cucumis sativus L.*) xylem sap by K-Edge TXRF-XANES
    - V. Groma, J. Osan, A. Alsecz, S. Török, F. Meirer, C. Streli, P. Wobrauschek, G. Falkenberg Trace element analysis of airport related aerosols using SR-TXRF

- Denver X-Ray Conference 07, Denver, USA; Poster (Best Poster Award) and Talk
    - F. Meirer, C.Streli, G. Pepponi, P. Wobrauschek, C. Horntrich, Mary Ann Zaitz, G. Falkenberg, Determination of the oxidation state of iron-contaminations on silicon wafer surfaces with K-edge TXRF XANES
- ICXOM 2007, Kyoto, Japan, Poster and Talk (Streli)
    - F. Meirer, G. Pepponi, C. Streli, P. Wobrauschek, P. Kregsamer, C. Horntrich, J. Broekaert, U. E. A. Fittschen, G. Falkenberg, Self-absorption effects in TXRF-XANES measurements – a parameter study
    - C. Streli, F. Meirer, G. Pepponi, P. Wobrauschek, G. Záray, U. E. A. Fittschen, J. Broekaert, G. Falkenberg, Synchrotron radiation induced TXRF for XANES application at HASYLAB, Beamline L
- Conference of the Austrian Physical Society 2007, Krems, Austria; Poster
    - F. Meirer, G. Pepponi, C. Streli, P. Wobrauschek, P. Kregsamer, C. Horntrich, J. Broekaert, U. E. A. Fittschen, G. Falkenberg, Self-absorption effects in TXRF-XANES measurements – a parameter study
- CSI XXXV, 2007, Xiamen, China, Talk (Streli)
    - C. Streli, F. Meirer, G. Pepponi, P. Wobrauschek, U. E. A. Fittschen, J. Broekaert, G. Záray, G. Falkenberg, Synchrotron radiation induced Total Reflection X-ray Fluorescence Analysis

The most recent work on the application of the grazing exit setup will be presented at the EXRS conference in 2008:

- EXRS 2008, Dubrovnik, Croatia; Poster
    - F. Meirer, G. Pepponi, C. Streli, P. Wobrauschek, N. Zoeger, A new Grazing-Exit-XRF setup at HASYLAB beamline L

Kurzfassung

Synchrotronstrahlungsinduzierte Totalreflexions-Röntgenfluoreszenzanalyse (engl. „SR-TXRF") ist eine mikroanalytische Methode, deren Nachweisgrenze für die meisten Elemente im Bereich von fg liegt. Die Verwendung einer Synchrotronstrahlungsquelle erlaubt die Kombination mit Absorptionsspektroskopie und dadurch eine Bestimmung des chemischen Zustandes eines spezifischen Elements, das nur in geringsten Mengen vorliegt. In der vorliegenden Arbeit wurde diese kombinierte Methode auf verschiedenste analytische Fragestellungen angewandt.

Für die Charakterisierung von Eisenkontaminationen auf Silizium-Wafer Oberflächen wurde eine Analyse der kantennahen Feinstruktur des Absorptionskoeffizienten (engl. „XANES") in Totalreflexionsgeometrie durchgeführt. Die chemische Elementanalyse von Kontaminationen auf Silizium-Wafern ist äußerst wichtig für die Halbleiterindustrie, um mögliche Quellen der Kontamination zu erkennen und zu beseitigen. Da eine solche Analyse mit Laborgeräten nicht möglich ist, war eine der Hauptaufgaben des Projekts die Machbarkeit einer XANES Studie an einem Synchrotron für die äußerst geringen Konzentrationen auf dem Substrat zu zeigen.

Um die Auswirkungen der Fein- und Ultrafeinstaubbelastung durch in der Luft gelöste Partikel (Aerosole) auf die menschliche Gesundheit, aber auch auf das globale Klima abschätzen zu können, ist die Kenntnis über Quellen und Transportmechanismen von Aerosolen unumgänglich. Eine Aerosol-Analyse sollte daher unter Berücksichtigung der Partikelgröße nicht nur Informationen über die in den Partikeln enthaltenen Elemente, sondern in speziellen Fällen auch über den chemischen Zustand eines spezifischen Elements liefern. Durch die Kombination von TXRF und XANES konnte sowohl der Oxidationszustandes des Eisens als auch die Elementzusammensetzung in Aerosolen bestimmt werden, die, nach Partikelgrößen aufgelöst, in verhältnismäßig kurzen Zeiträumen gesammelten wurden (was geringe Probenmengen zur Folge hat).

Die Anwendbarkeit der Methode wurde weiters für die Analyse des Oxidationszustandes von Arsen im Xylem von Gurken (Cucumis sativus L.) getestet. Die Toxizität von Arsen hängt maßgeblich vom Oxidationszustand und der chemischen Verbindung ab – darüber hinaus ist bekannt, dass manche Pflanzen den Oxidationszustand des Arsens ändern können. Die Problemstellung lautete daher, eine Änderung des Oxidationszustandes des Arsens im Xylem von Gurken, denen diese Fähigkeit nachgesagt wird, nachzuweisen. Obwohl die Arsen-

Kurzfassung

Konzentrationen im Bereich von 30 ng/mL lagen, konnte eine Reduktion von As(V) zu As(III) nachgewiesen werden.

Während der Messungen im Zuge dieser Arbeit, speziell bei der Analyse höher konzentrierter Proben (meistens Standards), wurden Dämpfungseffekte der Oszillationen in der Feinstruktur des Absorptionskoeffizienten beobachtet. Als Grund wurde ein Selbstabsorptionseffekt vermutet, bedingt durch die extrem kleinen Einfallswinkel der TXRF Geometrie. Dieser Effekt konnte durch Messungen und Simulationen, basierend auf einer einfachen Monte-Carlo Simulation, erfolgreich nachgewiesen und genauer untersucht werden. Diese Ergebnisse wurden durch weitere Experimente bestätigt, welche die Anwendbarkeit der zur TXRF invertierten Messgeometrie (engl. GE, „Grazing Exit" Geometrie) untersuchten.

Im Rahmen dieser Arbeit konnten die Anwendbarkeit, die Vielseitigkeit, sowie die Stärken und Schwächen der Kombination von TXRF und XANES Analyse zur Untersuchung geringster Probenmengen erfolgreich gezeigt werden.

Acknowledgement

I would like to thank everyone who supported me in writing this thesis.

This work would never have been possible without Eleonore and my parents Herta and Reinhard, who have supported and encouraged me throughout my studies.

Special thanks go to my supervisor Christina Streli for her guidance and the confidence she placed in me.

I would like to express my gratitude to Giancarlo Pepponi, Peter Wobrauschek, Norbert Zoeger and Peter Kregsamer for their generous help and support.

Many other people deserve to be thanked for their assistance and will be mentioned in no particular order:
Szabina Török, Janos Osan and Veronika Groma from the KFKI Atomic Energy research Institute, Budapest, Hungary
Jose Broekaert and Ursula Fittschen from the Department of Chemistry at the University of Hamburg, Germany
Gyula Záray and Victor Mihucz from the Department of Inorganic and Analytical Chemistry at the Eötvös University of Budapest, Hungary
Gerald Falkenberg from HASYLAB, Hamburg, Germany
Mary Ann Zaitz from IBM Microelectronics, Hopewell Junction, New York, USA
Lorenzo Lunelli from the FBK-irst, Povo (Trento), Italy

This work was supported by the Austrian Science Fund (FWF), project number P18299 and the European Commission, project number II-20042060.
Parts of this work were done within the frame of CRP (G4.20.02./1271) *"Unification of nuclear spectrometries: integrated techniques as a new tool for material research"*

Content

Introduction 1
   1.1 Presentation of the thesis 2

Synchrotron Radiation 3
   2.1 Introduction 3
   2.2 Basic properties 4
      2.2.1 Radiated power and its angular distribution 4
         2.2.1.1 Non-relativistic motion 6
         2.2.1.2 Relativistic motion 7
      2.2.2 Time structure and spectral distribution 13
         2.2.2.1 Brightness, Flux and Brilliance 18
         2.2.2.2 Polarization 20
      2.2.3 Summary 21
   2.3 Insertion devices 25
      2.3.1 Wavelength shifters 25
      2.3.2 Wigglers 26
      2.3.3 Undulators 27
      2.3.4 Free electron lasers 28

X-Ray Absorption Fine Structure and Total Reflection X-Ray Fluorescence Analysis .. 31
   3.1 X-Ray absorption and fluorescence 31
   3.2 X-Ray absorption fine structure 32
      3.2.1 Introduction 32
      3.2.2 Theory 33
      3.2.3 Experimental 38
   3.3 Total reflection X-Ray fluorescence analysis 40
      3.3.1 Introduction 40
      3.3.2 Theory 42
      3.3.3 Experimental 49
         3.3.3.1 Synchrotron radiation 49
         3.3.3.2 Quantification 50
         3.3.3.3 Grazing Exit Geometry 51

3.4 Summary 52

Silicon Wafer Surface Analysis of Fe contaminations 54

4.1 Introduction 54

4.2 Experimental 55

4.3 Results and discussion 56

4.4 Summary 63

Characterization of Atmospheric Aerosols 65

5.1 Introduction 65

    5.1.1 Analysis of airport related aerosol particles with high time resolution using SR-TXRF 66

    5.1.2 Characterization of atmospheric aerosols using SR-TXRF and Fe K-edge TXRF-XANES 67

5.2 Experimental 68

    5.2.1 Sampling in Budapest 68

    5.2.2 Sampling in Hamburg 69

    5.2.3 Measurements 70

        5.2.3.1 Aerosols collected with May impactor (at Budapest Airport) 72

        5.2.3.2 Aerosols collected with Berner impactor (in the city of Hamburg) 72

5.3 Results and discussion 73

    5.3.1 Results of the analysis of airport related aerosol particles 73

        5.3.1.1 Sample homogeneity 73

        5.3.1.2 Multi-element analysis 74

    5.3.2 Results of the characterization of atmospheric aerosols collected in the city of Hamburg 79

        5.3.2.1 Multi-element analysis 79

        5.3.2.2 Iron K-edge TXRF-XANES analysis 83

5.4 Summary 86

    5.4.1 Analysis of airport related aerosol particles with high time resolution using SR-TXRF 86

    5.4.2 Fe K-edge TXRF-XANES of atmospheric aerosols collected in the city of Hamburg 87

Analysis of Arsenic in Cucumber Xylem Sap 88

6.1 Introduction 88

6.1.1 Arsenic 88

6.1.2 SR-TXRF 90

6.1.3 Aims 91

6.2 Experimental 91

6.2.1 Sample preparation 91

6.2.2 Measurements 92

6.3 Results and discussion 94

6.4 Summary 100

Self Absorption Effects in TXRF-XANES analysis 101

7.1 Introduction 101

7.1.1 The self-absorption effect 101

7.1.2 Aims 103

7.2 Experimental 103

7.2.1 Sample preparation 103

7.2.2 TXRF-XANES measurements 104

7.2.3 Confocal microscopy 106

7.3 Development of a simple Monte-Carlo simulation 108

7.4 Results and discussion 112

7.4.1. Results of the measurements with the confocal microscope 112

7.4.2. TXRF-XANES measurements 115

7.4.3. Results of the Monte-Carlo simulations 118

7.5 Summary 121

Comparison of Grazing Exit and TXRF geometry for XANES analysis 123

8.1. Introduction 123

8.1.1 Grazing Exit (GE) geometry: 123

8.1.2 Aims 124

8.2 Experimental 124

8.2.1 Samples 125

8.2.2 Grazing exit measurements 126

8.3 Results and discussion 127

8.3.1 Angular scans 127

8.3.2 TXRF spectra 129

8.3.3 Area scans 131

    8.3.4 XANES analysis 132

    8.4 Summary 135

Concluding remarks 136

Bibliography 139

Chapter 1

Introduction

X-Ray fluorescence (XRF) spectrometry has been a powerful technique for elemental analysis for almost 100 years based on Moseley's well known law [1, 2] which relates "characteristic" fluorescence radiation to the atomic number of the emitting atom. Today composition analysis by measuring fluorescence spectra has become a routine technique utilized in a vast number of research areas ranging from material science to biomedical science. X-Ray fluorescence analysis is nondestructive, high precise and multi-elemental method to analyze most elements of the periodic table (Z>5 (B), most effective for Z>11 (Na)) with simple or even no sample preparation. Moreover, as the fluorescence intensity is proportional to the concentration of an element present in the sample, not only qualitative but also quantitative analysis is possible. As wavelength and energy are equivalent the fluorescence radiation can be evaluated in wavelength or energy dispersive mode. However, classical XRF analysis is not applicable for ultra-trace elemental analysis and susceptible to systematic errors due to sample matrix effects. A significant improvement regarding these points was achieved by the idea to use the effect of (external) total reflection of X-Rays to excite a sample applied on the surface the reflector. This technique was named after the effect of total reflection is now known as Total Reflection X-Ray fluorescence Analysis (TXRF). TXRF is a special technique of energy dispersive XRF analysis and is primarily used for chemical trace analysis today. Due to the special geometry which is required for a TXRF experiment and the effect of total reflection this method allows the detection of elements present in the pg range (and even in the fg region when Synchrotron radiation is used as excitation source).

Based on the Beer-Lambert law XRF spectrometry can also be used to analyze the absorption coefficient which is proportional to the emitted fluorescence. The measurement of the energy dependent absorption of X-Rays when penetrating through matter is the fundament of another extremely large field of research – X-Ray Absorption Spectroscopy (XAS). The investigation of the X-Ray Absorption Fine Structure (XAFS) is widely used to probe the physical and chemical structure of matter at an atomic scale. While diffraction techniques provide information about the atomic coordinates as a macroscopic average of a long range ordered periodical structure, XAFS gives information on the local atomic structure of a sample, i.e. the radial distribution and electronic states around a particular atomic species and short range ordered systems can be studied. A major requirement for XAFS analysis is an intense, tunable X-Ray source which explains why it is advantageous (or even indispensable) to use synchrotron radiation (SR) as excitation source. XAFS spectroscopy has advanced rapidly with the development of synchrotron radiation facilities and is now one of the major fields of application of this special radiation.

Synchrotron radiation designates the electromagnetic radiation emitted by a charged particle moving at relativistic speeds and following a curved trajectory. This radiation has several outstanding properties like high intensity, natural collimation and wide spectral range from the infrared to the X-ray region which make it an excellent radiation source for spectroscopy. Synchrotron radiation is nowadays of major importance as a radiation source and is used in various fields of research. Today a third generation of storage rings is exclusively dedicated for use as synchrotron radiation sources.

Synchrotron radiation also opened new possibilities in TXRF analysis extending the limits of detection to the fg range. Furthermore it enables the combination of TXRF and XAFS analysis which allows chemical speciation at an ultra trace level. Up to now only a few studies have focused on this very special combination – the present work is one of them.

1.1 Presentation of the thesis

The chapters 2 and 3 are dedicated to the theoretical fundamentals of Synchrotron radiation, X-Ray absorption spectroscopy and Total-Reflection X-Ray fluorescence analysis. The following chapters focus on special applications of Synchrotron radiation induced TXRF in combination with XAFS analysis. Chapter 4 deals with the analysis of contaminations found on the surface of a Silicon wafer which is used in semiconductor industry. The chapters 5 and 6 describe environmental applications, namely the analysis of atmospheric aerosols (fine and ultra fine dust) and the speciation of Arsenic in xylem of plants. Chapter 7 attends to the important fact that self absorption effects have to be considered for SR-TXRF-XAFS analysis when higher concentrated samples are investigated. To gain a deeper understanding of this geometry dependent effect chapter 8 presents results of measurements utilizing a geometry inverse to the TXRF setup – the grazing exit arrangement.

Chapters 4 to 8 include a short introduction describing the motivation and basic premises in each case – therefore these introductions include short repetitions of the fundamentals described in chapters 2 and 3. This was intended for better readability of the single chapters.

Chapter 2

Synchrotron Radiation

2.1 Introduction

A charged particle moving at relativistic speeds and following a curved trajectory emits electromagnetic radiation which is termed "synchrotron radiation" (SR) after its first visual observation in the General Electric (G.E.) 70 MeV Synchrotron in the year 1947 by F.R. Elder et al. [3]. Historical details about this discovery can be found in [4].

Shortly before this first detection of synchrotron radiation D. Iwanenko and I. Pomeranchuk (1944) [5] as well as J.P. Blewett (1946) [6] published articles dealing with energy losses of accelerated electrons due to SR. However, the theoretical basis for the understanding of synchrotron radiation goes back much further. In 1898 A. Liénard first presented the energy loss formula for charged particles moving on a circular path with relativistic speeds in his article entitled "The electric and magnetic field produced by an electric charge concentrated at a point and in arbitrary motion" [7]. G.A. Schott extended the theory and in 1912 published calculations for the frequency and angular distribution of the radiation as well as the polarization properties [8]. For more than thirty years the theory received no further interest but this changed dramatically with the construction of the first high energy accelerators. In 1945 J. Schwinger produced a paper in preprint form and presented the results as a 15-minute invited paper in 1946, at an American Physical Society meeting, under the title "Electron Radiation in High Energy Accelerators" (the abstract is published in [9]). Four years later he published this work in revised form [10]. The original paper was later transcribed by M.A. Furman in 1996 and published in [11]. In this work he in-depth pointed out the theory of the radiation from a high energy accelerated electron. The results of the following systematic investigations of the spectral distribution of the radiation carried out in the visible part of the spectrum showed good agreement with theory [12-14]. Spectral measurements for the UV and soft X-ray region were first carried out by Tomboulian and Hartman [15].

Due to its special properties SR was soon used as a radiation source for experiments. In the 1960's several SR facilities were set up on accelerators built initially for high energy physics. This "parasitic" use of the radiation was so successful that a second and recently a third generation of storage rings was built for dedicated use as synchrotron radiation sources. Synchrotron radiation is nowadays of major importance as a radiation source and used in various fields of research. Furthermore SR opened new possibilities in x-ray analysis as it represents an ideal x-ray source for most applications.

2.2 Basic properties

Synchrotron radiation has several outstanding properties making it an excellent radiation source for spectroscopy:

- High intensity
- Natural collimation (in the direction of flight of the emitting particles)
- Time structure (puls lengths down to 100ps)
- Wide spectral range from the infrared to the X-ray region
- Polarization
- Small source size (size and angular spread of the electron beam)

The biggest advantage for the use of SR is the absolute calculability of these properties. In the following a short summary about the theory of origin and characteristics of SR based on the work of J. Schwinger is given. A more detailed discussion on this topic can be found in [10, 16-22].

2.2.1 Radiated power and its angular distribution

Using the localized charge and current densities of a single charged particle

$$\rho(\bar{x},t) = e\delta^{(3)}(\bar{x}-\bar{r}(t)) \quad \text{and} \quad \bar{J}(\bar{x},t) = e\bar{v}(t)\delta^{(3)}(\bar{x}-\bar{r}(t)) \quad (2.1)$$

the inhomogeneous wave equations for the scalar potential Φ and the vector potential A

$$\bar{\nabla}^2 \Phi - \frac{1}{c^2}\frac{\partial^2 \Phi}{\partial t^2} = -\frac{\rho}{\varepsilon_0} \qquad \bar{\nabla}^2 \bar{A} - \frac{1}{c^2}\frac{\partial^2 \bar{A}}{\partial t^2} = -\frac{\bar{J}}{c^2 \varepsilon_0} \quad (2.2)$$

have as solution the Liénard-Wiechert potentials:

$$\bar{A}(\bar{x},t) = \frac{1}{4\pi\varepsilon_0 c}\left[\frac{e\bar{\beta}}{(1-\bar{\beta}\cdot\bar{n})R}\right]_{ret} \qquad \Phi(\bar{x},t) = \frac{1}{4\pi\varepsilon_0}\left[\frac{e}{(1-\bar{\beta}\cdot\bar{n})R}\right]_{ret} \quad (2.3)$$

The index ret means that the expression within the square brackets is to be evaluated at the retarded time $t = t' + R(t')/c$.

Chapter 2   Synchrotron Radiation

$\overline{\beta}$ is the normalized electron velocity

$$\overline{\beta} \equiv \overline{v}/c = \sqrt{1-(1/\gamma)^2} \quad \text{and} \quad \frac{dt}{dt'} = 1 + \frac{1}{c}\frac{dR(t')}{dt'} = 1 - \overline{\beta}\cdot\overline{n} \quad (2.4)$$

using the Lorentz-factor $\gamma = \left(\sqrt{1-\overline{\beta}^2}\right)^{-1}$. $\overline{n} = \overline{R}/R$ is a unit vector directing from the charge to the observer and R is the distance between them.

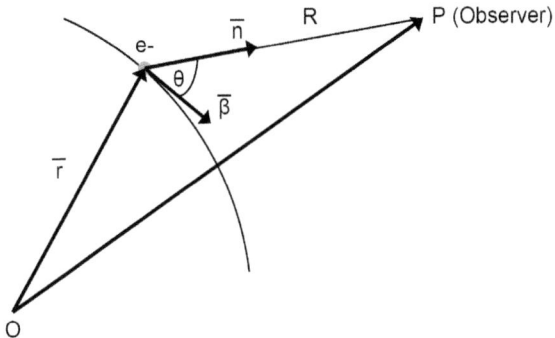

Figure 2.1: Sketch of the geometry for the moving charged particle and its observer

The electric and magnetic fields generated by the moving charge are related to the potentials by

$$\overline{E} = -\nabla\Phi - \frac{\partial \overline{A}}{\partial t} \qquad \overline{B} = \nabla\times\overline{A} \quad (2.5)$$

and are called Liénard-Wiechert fields:

$$\overline{E}(\overline{x},t) = \frac{e}{4\pi\varepsilon_0}\left[\frac{\overline{n}-\overline{\beta}}{\gamma^2(1-\overline{\beta}\cdot\overline{n})^3 R^2}\right]_{ret} + \frac{e}{4\pi\varepsilon_0 c}\left[\frac{\overline{n}\times(\overline{n}-\overline{\beta})\times\dot{\overline{\beta}}}{(1-\overline{\beta}\cdot\overline{n})^3 R}\right]_{ret} \quad (2.6a)$$

$$\overline{B}(\overline{x},t) = \frac{1}{c}\left[\overline{n}\times\overline{E}\right]_{ret} \quad (2.6b)$$

The first term of these fields is called "velocity field" and is independent of the acceleration. It is therefore not of concern in the following. The second term depends linearly on $\dot{\overline{\beta}}$ and is therefore called "acceleration field". The "acceleration fields" fall off like R^{-1} and E and B are normal to the radius vector. These properties are typical for radiation fields.

The radiated power is determined using the Poynting vector:

$$\overline{S} = \frac{1}{\mu_0}\overline{E}\times\overline{B} = \frac{1}{\mu_0 c}|\overline{E}|^2\overline{n} \quad (2.7)$$

Chapter 2   Synchrotron Radiation

This leads to the power radiated per unit area observed in the laboratory (time reference of the moving charge): $[\vec{S} \cdot \vec{n}]_{ret} (dt/dt')$. This expression has to be multiplied by R^2 to get the formula for the power radiated per unit solid angle (using equation (2.4)):

$$\frac{d^2 P}{d\Omega} = R^2 (\vec{S} \cdot \vec{n}) \left(\frac{dt}{dt'} \right) = (1 - \vec{n} \cdot \vec{\beta}) R^2 \vec{S} \cdot \vec{n} \qquad (2.8)$$

2.2.1.1 Non-relativistic motion

In a frame of reference where the velocity of the charged particle is low compared to that of light the acceleration field reduces to ($\vec{\beta} \approx \vec{0}$):

$$\vec{E}_{acc}(\vec{x},t) = \frac{e}{4\pi\varepsilon_0 c} \left[\frac{\vec{n} \times (\vec{n} \times \dot{\vec{\beta}})}{R} \right]_{ret} \qquad (2.9)$$

This formula shows, that the emitted radiation is polarized in the plane defined by $\dot{\vec{\beta}}$ and $\vec{n}$. For the radiated power one obtains:

$$\frac{d^2 P}{d\Omega} = R^2 \vec{S} \cdot \vec{n} = \frac{1}{\mu_0 c} \left| R\vec{E}_{acc} \right|^2 = \frac{e^2}{(4\pi)^2 \varepsilon_0 c} \left| \vec{n} \times (\vec{n} \times \dot{\vec{\beta}}) \right|^2 \qquad (2.10)$$

(using $1/c^2 = \varepsilon_0 \mu_0$)

Finally this expression can be written as a function of the angle between the acceleration $\dot{\vec{\beta}}$ and $\vec{n}$ (figure 1):

$$\frac{d^2 P}{d\Omega} = \frac{e^2}{(4\pi)^2 \varepsilon_0 c} \left| \dot{\vec{\beta}} \right|^2 \sin^2 \theta \qquad (2.11)$$

Integrating over all angles leads to Larmor's formula for a non-relativistic accelerated charge:

$$P = \frac{e^2}{6\pi\varepsilon_0 c} \left| \dot{\vec{\beta}} \right|^2 \qquad (2.12)$$

Figure 2.2 shows the radiation pattern of a charged particle according to the above equation. In the reference frame of the moving particle K' (or for a non-relativistic motion v<<c) the angular distribution of the radiation corresponds to that of Hertz's dipole.

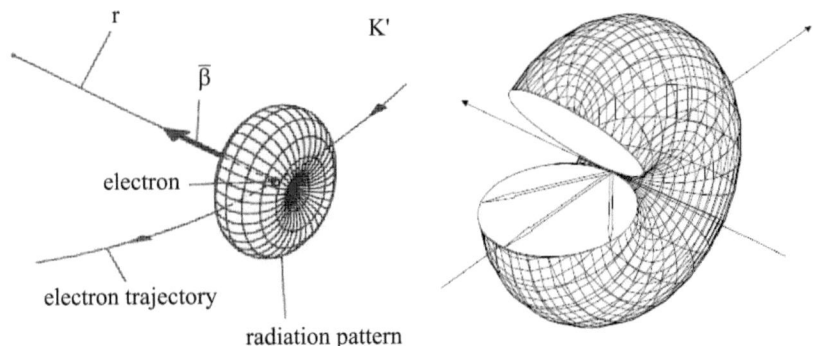

Figure 2.2 (adapted from [18] and [20]): Radiation pattern of a moved charged particle in its reference frame

2.2.1.2 Relativistic motion

If the electron moves with a velocity near the speed of light the formula for the radiated power per unit solid angle is given by:

$$\frac{d^2P}{d\Omega} = \frac{1}{\mu_o c}\left|\overline{RE}_{acc}\right|^2 = \frac{e^2}{(4\pi)^2 \varepsilon_0 c} \frac{\left|\overline{n}\times\left[(\overline{n}-\overline{\beta})\times\dot{\overline{\beta}}\right]\right|^2}{(1-\overline{n}\cdot\overline{\beta})^5} \quad (2.13)$$

This expression is dominated by the denominator $(1-\overline{n}\cdot\overline{\beta})^5 = (1-\beta\cos\theta)^5$ which shows, that for $\beta \rightarrow 1$ the emission is peaked in the direction of the velocity.

Integration over all angles delivers the relativistic generalization of Larmor's formula (Liénard 1898):

$$P = \frac{e^2}{6\pi\varepsilon_0 c}\gamma^6\left[(\dot{\overline{\beta}})^2 - (\overline{\beta}\times\dot{\overline{\beta}})^2\right] \quad (2.14)$$

Assuming $\overline{\beta} \perp \dot{\overline{\beta}}$ for a circular motion of the electron the radiated power per unit solid can be written as:

$$\frac{d^2P}{d\Omega} = \frac{e^2\left|\dot{\overline{\beta}}\right|^2}{(4\pi)^2 \varepsilon_0 c} \frac{1}{(1-\beta\cos\theta)^3}\left[1 - \frac{\sin^2\theta\cos^2\varphi}{\gamma^2(1-\beta\cos\theta)^2}\right] \quad (2.15)$$

(The coordinate system was chosen as shown in figure 2.3. The angles θ and φ define the direction of observation.)

Chapter 2   Synchrotron Radiation

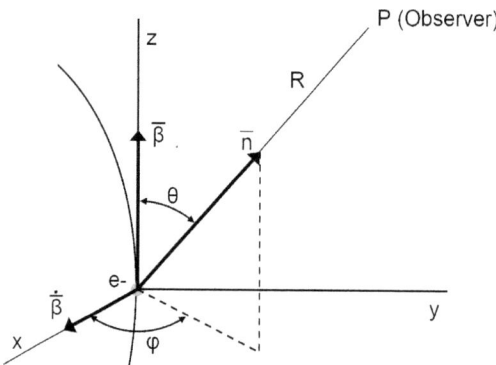

Figure 2.3 (as introduced by [19]): Coordinate system as defined for the derivation of equation (2.15)

When the electron velocity approaches the speed of light the emission pattern is sharply collimated forward.

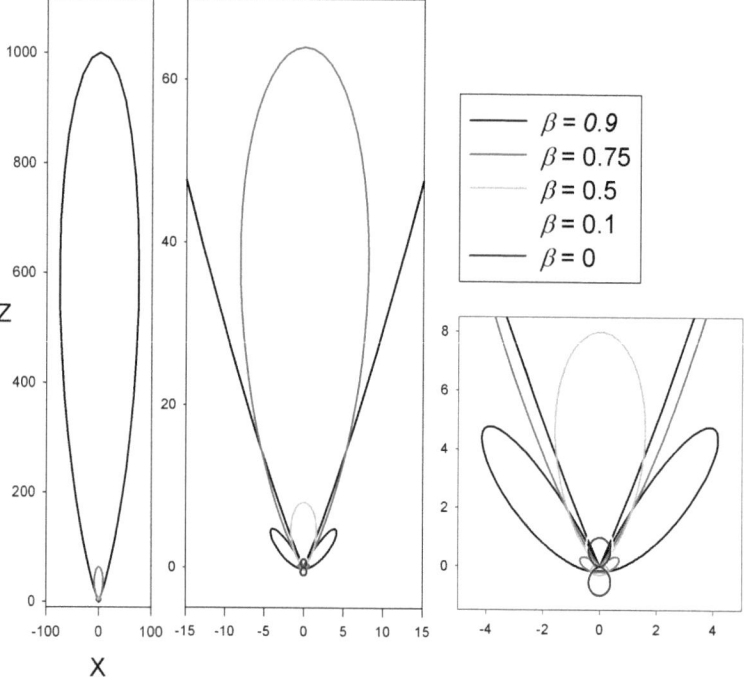

Figure 2.4: Radiation pattern in the x,z-plane ($\varphi = 0$) as a function of beta

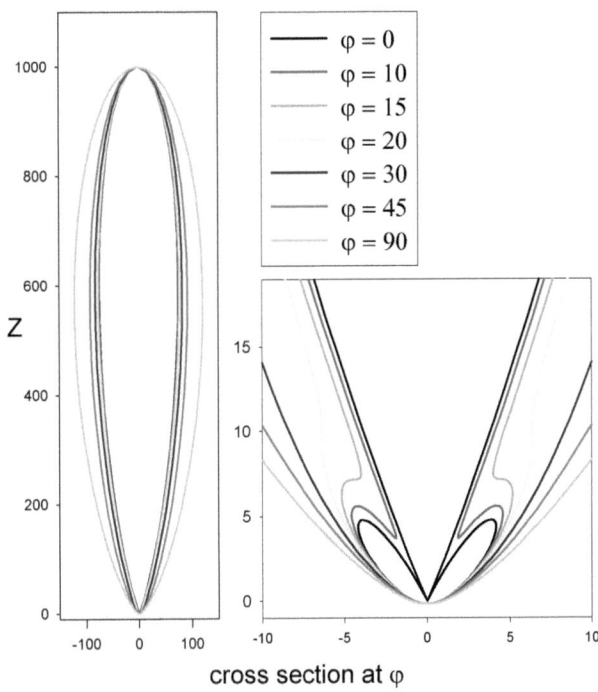

Figure 2.5: Radiation pattern plotted for different values of the angle phi (β=0.9)

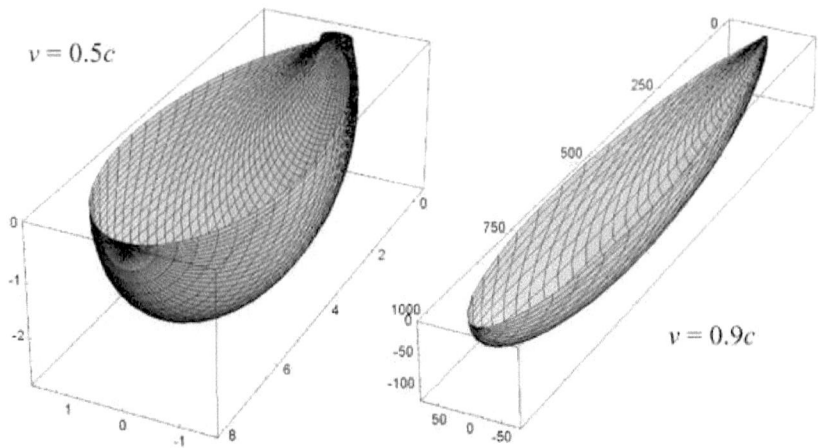

Figure 2.6 (source [23]): 3-dimensional plot of the radiation pattern for β=0.5 (left) and β=0.9 (right)

Chapter 2   Synchrotron Radiation

From (2.15) the root mean square (rms) angle of emission (mean angle between the direction of emission and that of the electron's motion) can be calculated to [10]:

$$\langle \theta^2 \rangle^{1/2} = m_0 c^2 / E = 1/\gamma, \quad \text{for } \gamma \gg 1 \tag{2.16}$$

(It will be shown later that this is the opening angle at the so called critical energy)
It is more instructive to derive this result considering the following:
The four-momentum p'_μ of a photon emitted perpendicular to the moving direction of the charged particle along the y-axis can be transferred to the observer's frame of reference K using the Lorentz transformation. According to (2.11), these photons represent the maximum intensity.

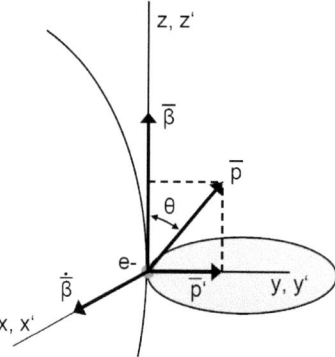

Figure 2.7: For positive values of y the intensity distribution of Hertz's dipole (frame of reference of the particle K') in the x,y-plane is shown in blue.

$$p'_\mu = \begin{pmatrix} E'/c \\ 0 \\ p'_y \\ 0 \end{pmatrix} \quad \text{and } \bar{v} \parallel \bar{z} \text{ hence } L = \begin{pmatrix} \gamma & 0 & 0 & \beta\gamma \\ 0 & 1 & 0 & 0 \\ 0 & 0 & 1 & 0 \\ \beta\gamma & 0 & 0 & \gamma \end{pmatrix} \tag{2.17}$$

($p'_y = E'/c$ is the momentum of a photon).
With (2.17) one obtains:

$$p_\mu = L p'_\mu = \begin{pmatrix} \gamma E'/c \\ 0 \\ p'_y \\ \beta\gamma E'/c \end{pmatrix} \tag{2.18}$$

The angle θ can now be calculated to:

$$\tan(\theta) = \frac{p_y}{p_z} = \frac{-p'_y}{-\beta\gamma E'/c} = \frac{1}{\beta\gamma} \Rightarrow \theta \approx \frac{1}{\gamma}, \text{ for } \gamma \gg 1 \tag{2.19}$$

This is consistent with (2.16). (Of course the same result is achieved for a photon emitted along the x-axis.)

To give an example: if the energy E is assumed to be 1 GeV γ is then 1957 (for electrons) and θ becomes 0.5mrad = 0.03°.

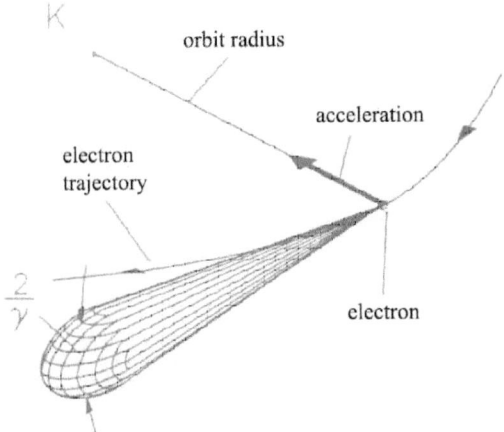

Figure 2.8 (adapted from [24]): In the frame of the observer the radiation is emitted in a very narrow cone with an opening angle of 2/γ.

By integrating (2.15) over the whole solid angle or directly from the relativistic generalization of Larmor's formula (using $\vec{\beta} \perp \dot{\vec{\beta}}$) one obtains the total instantaneous power radiated by one electron:

$$P = \frac{e^2}{6\pi\varepsilon_0 c} |\dot{\vec{\beta}}|^2 \gamma^4 = \frac{e^2 c}{6\pi\varepsilon_0} \frac{\beta^4}{r^2} \gamma^4 = \frac{e^2}{6\pi\varepsilon_0 m_0^2 c^3} \left|\frac{d\overline{p}}{dt}\right|^2 \gamma^2 = \frac{e^4 \beta^2}{6\pi\varepsilon_0 m_0^4 c^5} E^2 B^2 \quad (2.20)$$

with $|\dot{\vec{v}}| = v^2/r$ (centripetal acc.) in $|\dot{\vec{\beta}}|^2 = |\dot{\vec{v}}|^2/c^2$ to get the 2nd and $\dot{\vec{v}} = \frac{1}{\gamma m_0} \frac{d\overline{p}}{dt}$ to derive the 3rd expression.

The last term is the radiated power in terms of the energy $E = \gamma m_0 c$ and the magnetic field of the bending dipole B used to accelerate the electron according to the Lorentz equation:

$$\left|\frac{d\overline{p}}{dt}\right| = \left|e\left(\overline{E} + \frac{\overline{v} \times \overline{B}}{c}\right)\right| = \frac{evB}{c} \quad \text{using } \overline{E} = \overline{0}, \overline{B} \perp \overline{v} \text{ (bending magnet)} \quad (2.21)$$

Chapter 2   Synchrotron Radiation

It can be seen from the above expressions that the power radiated by a relativistic charge is proportional to $1/r^2$ (r is the bending radius) and B^2, the magnetic field of the bending dipole. Furthermore it has a strong dependence on the rest mass ($1/m_0^4$). This is the reason why it is advantageous to use electrons instead of protons in synchrotrons. Comparing the radiated power of an electron with that of a proton with the same energy using

$$m_e c^2 = 0.511 MeV \quad m_p c^2 = 938.3 MeV \quad (2.22)$$

one obtains:

$$\frac{P_e}{P_p} = \left(\frac{m_p c^2}{m_e c^2}\right)^4 = 1.13E13 \quad (2.23)$$

In the following some basic synchrotron radiation relationships are given according to (2.20): In the time T_b spent in the bending magnets the particle loses the energy U_0:

$$U_0 = \int P dt = PT_b = P\frac{2\pi r}{c} = \frac{e^2}{3\varepsilon_0} \frac{\beta^4 \gamma^4}{r} \quad (2.24)$$

(below $\beta \approx 1$)

Therefore the energy loss per turn (per electron) is:

$$U_0[keV] = \frac{e^2 \gamma^4}{3\varepsilon_0 r} = 88.46 \frac{E[GeV]^4}{r[m]} \quad (2.25)$$

The power radiated by a beam of average current I_b:

$$N_{tot} = \frac{I_b \cdot T_{rev}}{e} \quad P[kW] = \frac{U_0 N_{tot}}{T_{rev}} = \frac{e\gamma^4}{3\varepsilon_0 r} I_b = 88.46 \frac{E[GeV]^4 I[A]}{r[m]} \quad (2.26)$$

N_{tot} ... total number of electrons circulating
T_{rev} ... time for one revolution of the electrons in the storage ring

This power loss has to be compensated by the RF (radio-frequency) system. The radiation power emitted by an electron beam in a storage ring is very high. Introducing the parameters of e.g. DORIS III [25] leads to:

Energy [GeV] = 4.45

$$P[kW] = 88.46 \frac{(4.45)^4 \cdot 0.14}{12.18} \cong 400$$

Current [A] = 0.14

Bending radius [m] = 12.18

$$U_0[keV] = 88.46 \frac{(4.45)^4}{12.18} \cong 2850$$

The real value for the energy loss per turn (U_0) is 3466 keV [25] due to the installation of insertion devices (wigglers and undulators) different from bending magnets.

So far we have discussed two of the main features of synchrotron radiation, namely the high radiation power (or intensity) and the natural collimation in the direction of flight of the emitting particles. In the next chapter will focus on the time structure and the spectral range of synchrotron radiation.

2.2.2 Time structure and spectral distribution

Due to the strong collimation in the forward direction ($\theta = 1/\gamma$) of the emitting particle the observer detects only an extremely short-time pulse of synchrotron light.

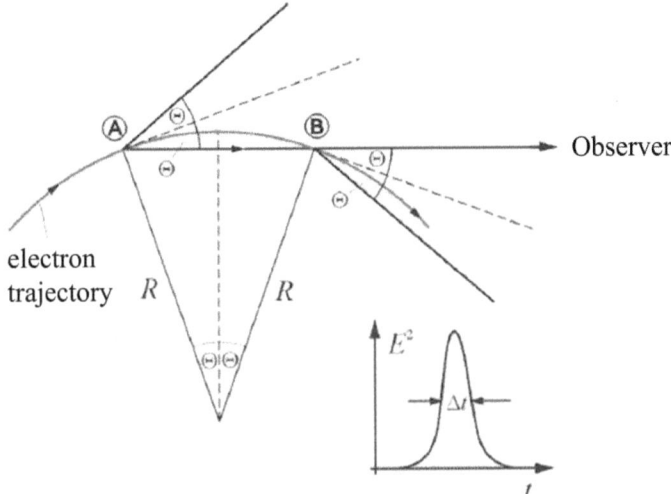

Figure 2.9 (adapted from [17]): Origin of the electromagnetic pulse generated by a relativistic electron moving on a circular orbit

The length of the pulse is defined by the difference in time of flight of the electron (on its trajectory) and the photon between A and B:

$$\Delta t = t_e - t_p = \frac{2R\Theta}{c\beta} - \frac{2R\sin\Theta}{c} \qquad (2.27)$$

Using the Tylor series for $\sin\Theta$ and $\gamma\beta \approx \gamma - 1/2\gamma$ one obtains:

$$\Delta t = \frac{2R}{c}\left(\frac{\Theta}{\beta} - \Theta + \frac{\Theta^3}{3!} - ...\right) = \frac{2R}{c}\left(\frac{1}{\gamma - 1/2\gamma} - \frac{1}{\gamma} + \frac{1}{6\gamma^3}\right) \qquad (2.28)$$

Chapter 2   Synchrotron Radiation

With

$$\left(\frac{1}{\gamma - 1/2\gamma}\right) = \frac{1}{\gamma}\frac{1}{1 - 1/2\gamma^2} \approx \frac{1}{\gamma}\left(1 + \frac{1}{2\gamma^2}\right) = \frac{1}{\gamma} + \frac{1}{2\gamma^3} \tag{2.29}$$

(2.28) becomes:

$$\Delta t \approx \frac{2R}{c}\left(\frac{1}{\gamma} + \frac{1}{2\gamma^2} - \frac{1}{\gamma} + \frac{1}{6\gamma^3}\right) = \frac{4R}{3c\gamma^3} \tag{2.30}$$

If this short pulse is transformed to frequency space by means of a Fourier transformation the resulting frequency spectrum is very broad with a typical frequency:

$$\omega_{typ} = \frac{2\pi}{\Delta t} = \frac{3\pi}{2}\frac{c\gamma^3}{R} \tag{2.31}$$

Commonly the critical frequency is defined which separates the frequency spectrum in two regions of equal radiation power:

$$\omega_c = \frac{\omega_{typ}}{\pi} = \frac{3}{2}\frac{c\gamma^3}{R} = \frac{3}{2}\gamma^3\omega_0 \tag{2.32}$$

For circular motion c/R is the angular frequency of rotation ω_0. Formula (2.32) shows that for E>>mc² a relativistic particle emits a broad spectrum of frequencies up to γ^3 times ω_0.

The frequency spectrum was calculated by Schwinger [10] and in the following just a short summary of the derivation is given. Details can be found in [19].

In general the energy received by an observer (i.e. expressed in the observer's time) per unit solid angle at the source is:

$$\frac{d^2W}{d\Omega} = \int_{-\infty}^{\infty}\frac{d^2P}{d\Omega}dt = c\varepsilon_0\int_{-\infty}^{\infty}|R\overline{E}(t)|^2\,dt \tag{2.33}$$

Where $\overline{E}$ is the electric field (2.6a). Now the Fourier transformation is used to move to the frequency space:

$$\frac{d^2W}{d\Omega} = 2c\varepsilon_0\int_0^{\infty}|R\overline{E}(\omega)|^2\,d\omega \tag{2.34}$$

The frequency and angular distribution of the energy received by an observer can now be expressed with:

$$\frac{d^3I}{d\Omega d\omega} = 2\varepsilon_0 cR^2\left|\hat{\overline{E}}(\omega)\right|^2 \tag{2.35}$$

Assuming the observer in the far field (n and R constant; see figure 2.1) and neglecting the velocity fields one obtains the so-called "Radiation Integral":

Chapter 2   Synchrotron Radiation

$$\frac{d^3I}{d\Omega d\omega} = \frac{e^2}{4\pi\varepsilon_0 4\pi^2 c} \left| \int_{-\infty}^{\infty} \frac{\vec{n} \times [(\vec{n} - \vec{\beta}) \times \dot{\vec{\beta}}]}{(1 - \vec{n} \cdot \vec{\beta})^2} e^{i\omega(t - \vec{n} \cdot \vec{r}(t)/c)} dt \right|^2 \quad (2.36)$$

This expression can be simplified to (see [19]):

$$\frac{d^3I}{d\Omega d\omega} = \frac{e^2 \omega^2}{4\pi\varepsilon_0 4\pi^2 c} \left| \int_{-\infty}^{\infty} \vec{n} \times (\vec{n} \times \vec{\beta}) e^{i\omega(t - \vec{n} \cdot \vec{r}(t)/c)} dt \right|^2 \quad (2.37)$$

The distribution of energy in frequency and angle can now be determined using the radiation integral (2.37). To solve this integral Airy integrals or the modified Bessel functions are required.

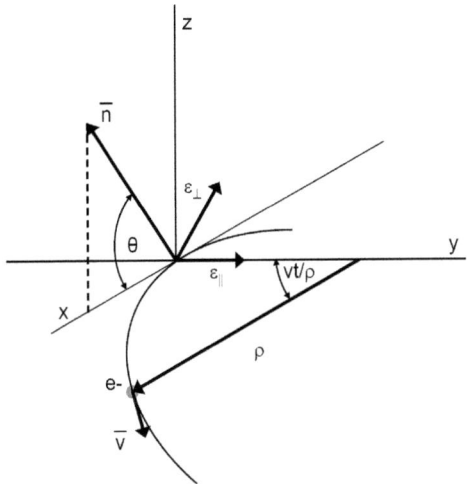

Figure 2.10 (as introduced by [19]): Coordinate system showing the geometry assumed in equation (2.38)

For the simple case of an electron in a bending magnet the trajectory of the arc of circumference can be written for small angles (according to figure 2.10):

$$\vec{r}(t) = \left(\rho \left(1 - \cos\frac{\beta c}{\rho} t \right), \quad \rho \sin\frac{\beta c}{\rho} t, \quad 0 \right) \quad (2.38)$$

At t = 0 the particle is at the origin of coordinates. The vector part of the integrand can be expressed with:

$$\vec{n} \times (\vec{n} \times \vec{\beta}) = \beta \left[-\vec{\varepsilon}_{\parallel} \sin\left(\frac{\beta c t}{\rho}\right) + \vec{\varepsilon}_{\perp} \cos\left(\frac{\beta c t}{\rho}\right) \sin\theta \right] \quad (2.39)$$

Here $\vec{\varepsilon}_{\parallel}$ is a unit vector in the y-direction, corresponding to the polarization in the orbital plane. $\vec{\varepsilon}_{\perp} = n \times \vec{\varepsilon}_{\parallel}$ is the orthogonal polarization vector corresponding approximately to the

15

Chapter 2   Synchrotron Radiation

polarization perpendicular to the orbital plane (for small values of θ). Finally the phase factor becomes:

$$\omega\left(t - \frac{\bar{n} \cdot \bar{r}(t)}{c}\right) = \omega\left(t - \frac{\rho}{c}\sin\left(\frac{\beta ct}{\rho}\right)\cos\theta\right) \cong \frac{\omega}{2}\left[\left(\frac{1}{\gamma^2} + \theta^2\right)t + \frac{c^2}{3\rho^2}t^3\right] \quad (2.40)$$

(for small angles θ, short times around t = 0 and β ≈ 1.)
Substituting (2.38) - (2.40) into the radiation integral (2.37) and introducing

$$\xi = \frac{\rho\omega}{3c\gamma^3}\left(1 + \gamma^2\theta^2\right)^{3/2} \quad (2.41)$$

one obtains:

$$\frac{d^3 I}{d\Omega d\omega} = \frac{e^2}{16\pi^3\varepsilon_0 c}\left(\frac{2\omega\rho}{3c\gamma^2}\right)^2 \left(1 + \gamma^2\theta^2\right)^2 \left[K_{2/3}^2(\xi) + \frac{\gamma^2\theta^2}{1+\gamma^2\theta^2}K_{1/3}^2(\xi)\right] \quad (2.42)$$

where $K_{2/3}$ and $K_{1/3}$ are the modified Bessel functions of fractional order. Formula (2.42) determines the energy radiated per unit frequency interval per unit solid angle. The first term in the square brackets corresponds to the radiation polarized in the orbital plane, whereas the second term is related to the radiation polarized perpendicular to that plane. The properties of the modified Bessel functions show that the radiation intensity is negligible for ξ >> 1. From (2.41) it can bee seen that this will be the case at large angles θ. If the frequency is increased the critical angle θ_c beyond which only negligible radiation will be emitted is decreased. The radiation is mainly restricted to the plane of motion (the higher the frequency – the larger the confinement). However, if ω gets too large ξ will be large at all angles and negligible total energy will be radiated at that frequency. As before where the critical angle was defined a critical frequency ω_c can be introduced beyond which there is negligible radiation at any angle:

$$\omega_c = \frac{3}{2}\frac{c}{\rho}\gamma^3 \quad (2.44)$$

This result is consistent with (2.32). (The numerical factor 3/2 is chosen so that the line $\omega/\omega_c = 1$ divides the area under the curve $S(\omega/\omega_c)$ into two equal parts - see figure 2.12). The critical angle can now be expressed with:

$$\theta_c = \frac{1}{\gamma}\left(\frac{\omega_c}{\omega}\right)^{1/3} \quad (2.45)$$

Figure 2.11 shows qualitatively the angular distribution for different frequencies (larger, equal and smaller than ω_c). For frequencies comparable to the critical frequency the radiation is confined to angles of the order of γ^{-1}. For much smaller (larger) frequencies, the angular spread is larger (smaller).

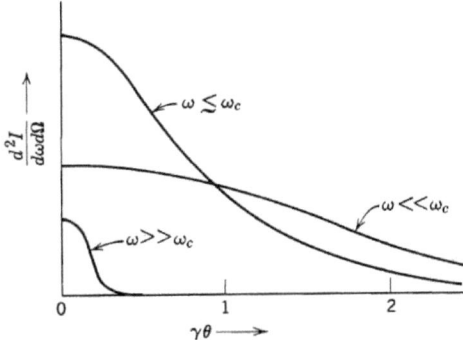

Figure 2.11 (source [19]): The differential frequency spectrum of synchrotron radiation displayed as a function of angle. The natural unit of angle $\gamma\theta$ is used.

By integrating (2.42) over all angles the frequency distribution of the radiated energy can be obtained:

$$\frac{dI}{d\omega} = \iint_{4\pi} \frac{d^3I}{d\omega d\Omega} d\Omega = \frac{\sqrt{3}e^2}{4\pi\varepsilon_0 c} \gamma \frac{\omega}{\omega_c} \int_{\omega/\omega_c}^{\infty} K_{5/3}(x) dx \qquad (2.46)$$

For the low- ($\omega<<\omega_c$) and high-frequency ($\omega>>\omega_c$) limits this expression reduces to:

$$\frac{dI}{d\omega} \approx \frac{e^2}{4\pi\varepsilon_0 c} \left(\frac{\omega\rho}{c}\right)^{1/3} \text{ for } \omega<<\omega_c \qquad (2.47)$$

and $\quad \dfrac{dI}{d\omega} \approx \sqrt{\dfrac{3\pi}{2}} \dfrac{e^2}{4\pi\varepsilon_0 c} \gamma \left(\dfrac{\omega}{\omega_c}\right)^{1/2} e^{-\omega/\omega_c}$ for $\omega>>\omega_c$ $\quad (2.48)$

Formula (2.47) shows that the spectrum increases as $\omega^{1/3}$ for $\omega<<\omega_c$ which results in a very broad, flat spectrum at frequencies below ω_c.

The radiation represented by (2.42) and (2.46) is called synchrotron radiation. Its spectral distribution depends only on the particle energy, the critical frequency and the purely mathematical Bessel functions. If the spectral distribution is normalized to the critical frequency, it does no longer depend on the particle energy and can therefore be represented by a universal distribution $S(\omega/\omega_c)$:

$$S(\frac{\omega}{\omega_c}) = \frac{9\sqrt{3}}{8\pi} \frac{\omega}{\omega_c} \int_{\omega/\omega_c}^{\infty} K_{5/3}(x) dx \quad \text{with: } \int_0^{\infty} S(x) dx = 1 \text{ and: } \int_0^{1} S(x) dx = \frac{1}{2} \qquad (2.49)$$

The last normalization condition shows, that the critical frequency divides the spectrum into two parts of equal power. With (2.49) expression (2.46) can now be written as:

$$\frac{dI}{d\omega} = \frac{\sqrt{3}e^2\gamma}{4\pi\varepsilon_0 c} \frac{\omega}{\omega_c} \int_{\omega/\omega_c}^{\infty} K_{5/3}(x)dx = \frac{2e^2\gamma}{9\varepsilon_0 c} S(\frac{\omega}{\omega_c}) \qquad (2.50)$$

In figure 2.12 the approximations of the universal function for the high ($\omega \gg \omega_c$) and low ($\omega \ll \omega_c$) frequency limits are given which are consistent with (2.47) and (2.48) and result from the asymptotic behaviour of the Bessel function.

Eventually, if (2.50) is integrated over all frequencies the well known relation for the total radiated power (2.20) can be found (starting from (2.24) and using (2.44) and (2.49)):

$$P = \frac{U_0}{T_b} = \frac{1}{T_b} \int_0^{\infty} \frac{dI}{d\omega} d\omega = \frac{1}{T_b} \frac{2e^2\gamma}{9\varepsilon_0 c} \omega_c \int_0^{\infty} S(x)dx = \frac{c}{2\pi\rho} \frac{2e^2\gamma}{9\varepsilon_0 c} \frac{3c\gamma^3}{2\rho} \cdot 1 = \frac{e^2 c}{6\varepsilon_0 \pi} \frac{\gamma^4}{\rho^2} \qquad (2.51)$$

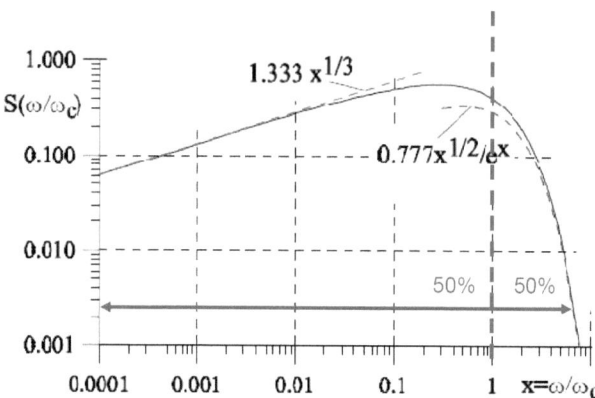

Figure 2.12 (adapted from [20]): Normalized universal function $S(\omega/\omega_c)$ of the synchrotron radiation spectrum. The frequency integral extended up to the critical frequency contains half of the total energy radiated, the peak occurs approximately at $0.3\omega_c$.

2.2.2.1 Brightness, Flux and Brilliance

For practical reasons it is convenient to define the critical energy and the critical wavelength:

$$\varepsilon_c = \hbar \omega_c = \frac{3}{2} \frac{\hbar c}{\rho} \gamma^3 \qquad (2.52)$$

$$\lambda_c = \frac{2\pi c}{\omega_c} = \frac{4\pi}{3} \frac{\rho}{\gamma^3} \qquad (2.53)$$

For electrons, the critical energy and the critical wavelength in practical units can be calculated with:

$$\varepsilon_c [keV] = 2.218 \frac{E[GeV]^3}{\rho[m]} = 0.665 \cdot E[GeV]^2 \cdot B[T] \qquad (2.54)$$

Chapter 2 Synchrotron Radiation

$$\lambda_c[A] = 5.59 \frac{\rho[m]}{E[GeV]^3} = \frac{18.6}{E[GeV]^2 \cdot B[T]} \quad (2.55)$$

Formula (2.42) gives the energy radiated per unit frequency interval per unit solid angle during the passage of a single particle. In the case of a beam of particles the total energy per second, i.e. the radiated power (in Watts) is proportional to the number of particles that pass the observer per second:

$$\frac{d^3P}{d\Omega d\omega} = \frac{d^3I}{d\Omega d\omega} \frac{I_b}{e} \quad (2.56)$$

where I_b is the average beam current in Ampere. Using practical units and the photon energy ε expressed in eV one obtains (in [Watts/mrad²/eV]):

$$\frac{d^3P}{d\Omega d\varepsilon} = 2.124 \cdot 10^{-3} \, E^2[GeV] \, I_b[A] \left(\frac{\varepsilon}{\varepsilon_c}\right)^2 (1+\gamma^2\theta^2)^2 \left[K_{2/3}^2(\xi) + \frac{\gamma^2\theta^2}{1+\gamma^2\theta^2} K_{1/3}^2(\xi)\right] \quad (2.57)$$

where (2.41) is expressed in terms of ε:

$$\xi = \frac{\rho\omega}{3c\gamma^3}(1+\gamma^2\theta^2)^{3/2} = \frac{\omega}{2\omega_c}(1+\gamma^2\theta^2)^{3/2} = \frac{\lambda_c}{2\lambda}(1+\gamma^2\theta^2)^{3/2} = \frac{\varepsilon}{2\varepsilon_c}(1+\gamma^2\theta^2)^{3/2} \quad (2.58)$$

It is also common to express SR intensities in terms of the number of photons per second. This is obtained by dividing the power in a given frequency interval by the appropriate photon energy $\hbar\omega$. Alternatively the power divided by $\hbar$ gives the number of photons per second per unit relative bandwidth:

$$\frac{d^3N}{d\Omega \, d\omega/\omega} = \frac{d^3P}{d\Omega d\omega} \frac{1}{\hbar} \quad \text{in [photons/sec/mrad²/0.1\% bandwidth/mA]} \quad (2.59)$$

This photon number is called "spectral brightness". Eventually, the number of photons per second in 1mrad² of solid angle, 0.1% bandwidth, and a beam current of 1mA in terms of the photon wavelength is given by (in [photons/sec/mrad²/0.1% bandwidth/mA]):

$$\frac{d^4N}{dt d\Omega d\lambda/\lambda} = 3.46 \cdot 10^3 \, \gamma^2 \left(\frac{\lambda_c}{\lambda}\right)^2 (1+\gamma^2\theta^2)^2 \left[K_{2/3}^2(\xi) + \frac{\gamma^2\theta^2}{1+\gamma^2\theta^2} K_{1/3}^2(\xi)\right] \quad (2.60)$$

(ξ is defined by (2.57)).
In the orbital plane ($\theta = 0$) at the critical wavelength ($\lambda = \lambda_c \Rightarrow \xi = 1/2$) (2.59) reduces to:

$$\left(\frac{d^4N}{dt d\Omega d\lambda/\lambda}\right)_{\theta=0, \lambda=\lambda_c} = 5.04 \cdot 10^3 \, \gamma^2 \quad (2.61)$$

(Using $K_{2/3}(0.5) = 1.206$)
If (2.59) is integrated over all angles the "spectral flux" is obtained, which is the photon number emitted per unit time, per unit band width, and a beam current of 1mA:

19

$$\frac{d^2N}{dt\,d\lambda/\lambda} \text{ in [photons/sec/0.1\% bandwidth/mA]} \qquad (2.62)$$

Finally the "total flux" in [photons/s/mA] can be calculated by integrating the spectral flux over all wavelengths.

Previously it was assumed that the radiation was emitted from an electron in ideal circular motion. In practice, the electron in the storage ring moves in an equilibrium orbit with some amount of fluctuation in space and angle. The flux is determined only by the electron energy and is not related to the size and angular spread of the electron beam. According to the definition of the brightness (2.59), which is the flux represented for unit solid angle, the angular divergence in reality is given by the convolution of the angular width of the electron beam with the intrinsic angular width of the radiation. The "spectral brilliance" is the brightness divided by the size of the radiation source (i.e. the cross section of the electron beam):

$$\frac{d^5N}{dt\,d\Omega\,dS\,d\lambda/\lambda} \text{ in [photons/sec/mrad}^2\text{/mm}^2\text{/0.1\% bandwidth/mA]} \qquad (2.63)$$

where S represents the area of the radiation source, being related to the size and angular spread of the electron beam.

2.2.2.2 Polarization

As already mentioned the two terms in the square brackets in formula (2.42) are associated with the intensities in the two directions of the polarization, I_P and I_N. P and N identify the contribution with the electric vector parallel (i.e. in the orbit plane) and normal to the acceleration direction. Figure 2.13 shows the angular distribution of the two components. In the orbital plane ($\theta = 0$), the radiation is purely linearly polarized ($I_P = 1$ while $I_N = 0$). The normal component I_N has small peaks above ($\theta > 0$) and below ($\theta < 0$) the orbital plane.

The degree of linear polarization P_l of the radiation is dependent on the angle θ and is shown in figure 2.13b:

$$P_l = \frac{I_P - I_N}{I_P + I_N} = \frac{K_{2/3}^2(\xi) - \dfrac{\gamma^2\theta^2}{1+\gamma^2\theta^2}K_{1/3}^2(\xi)}{K_{2/3}^2(\xi) + \dfrac{\gamma^2\theta^2}{1+\gamma^2\theta^2}K_{1/3}^2(\xi)} \qquad (2.64)$$

By integrating of (2.42) over all frequencies we get the angular distribution of the energy radiated:

$$\frac{d^2I}{d\Omega} = \int_0^\infty \frac{d^3I}{d\omega d\Omega}d\omega = \frac{7e^2\gamma^5}{64\pi\varepsilon_0\rho}\frac{1}{(1+\gamma^2\theta^2)^{5/2}}\left[1+\frac{5}{7}\frac{\gamma^2\theta^2}{(1+\gamma^2\theta^2)}\right] \qquad (2.65)$$

Integration of (2.65) over all angles shows that the intensity polarized parallel to the plane of the orbit is about 7 times larger than the intensity polarized perpendicular to the orbital plane.

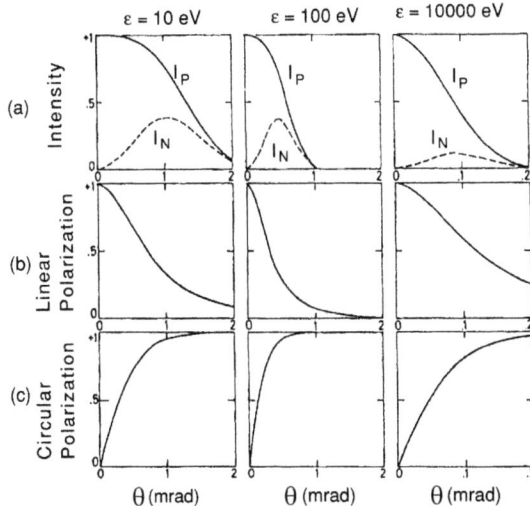

Figure 2.13 (adapted from [26]): a): Angular distribution of the intensity components with the electric vector parallel (I_P) and normal (I_N) to the orbital plane. b) and c): Linear and circular polarization (from decomposition into left (I_L) and right (I_R) hand circularly polarized components) for a storage ring with $\rho = 12.12$m and an energy of $E = 3.5$GeV calculated for three photon energies.

2.2.3 Summary

In the previous chapters the basic properties of synchrotron radiation, which were introduced at the beginning have been derived. In the following a short summary of the calculations is given:

High intensity

In section 2.2.1.2 the radiated power per unit solid angle and its angular distribution assuming circular motion of a relativistic electron was given (2.15). From (2.15) formula (2.26) was derived, which showed, that the radiation power emitted by an electron beam in a storage ring is very high: it has a strong dependence on the electron energy (E^4), is indirectly proportional to the diameter of the storage ring ($1/r^2$) and directly proportional to the number of electrons. Formula (2.51) in section 2.2.2 shows, that the same result is obtained taking into account the spectral distribution of the synchrotron radiation.

Chapter 2   Synchrotron Radiation

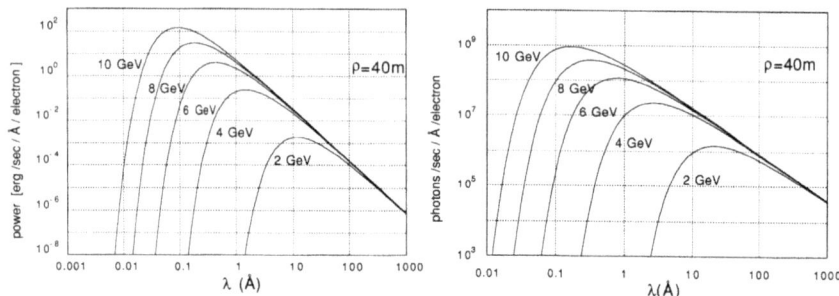

Figure 2.14 (source [22]): Spectral distributions of SR represented by left: the radiation power and right: the photon numbers ("flux") as given by formula (2.46) at various electron energies for an orbit radius of 40m. It can be seen that the curves show no dependence on the energy at longer wavelengths ($\lambda \gg \lambda_c$) which was expected from formula (2.47). The critical wavelengths λ_c for the different energies are determined by (2.53) and the maxima of the distributions occur at approximately $3\lambda_c$.

Natural collimation

The collimation of the synchrotron radiation is described by formula (2.15). In the frame of the observer the radiation is strongly collimated in the direction of flight of the emitting particles. The radiation pattern forms a very narrow cone with an opening angle of $2/\gamma$ (formulas (2.16) and (2.19)). The angular distribution for different frequencies is given by formula (2.42). For energies comparable to the critical energy the radiation is confined to angles of the order of γ^{-1}. For smaller wavelengths, the angular spread is smaller (and vice versa). This is - of course - consistent with (2.15).

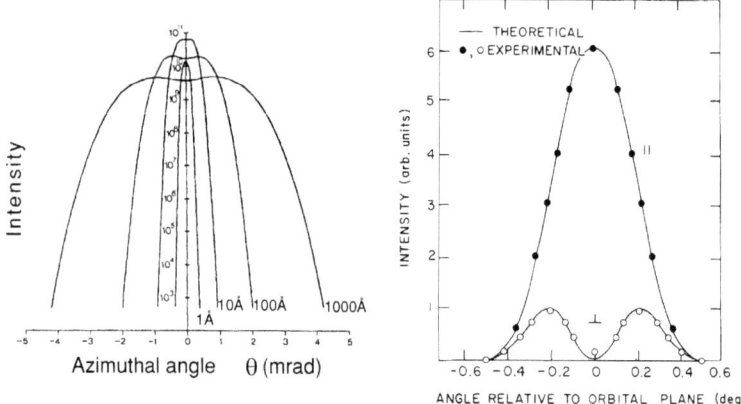

Figure 2.15: Left (adapted from [27]): Angular distribution of SR. The photon number is shown as a function of the angle θ perpendicular to the orbital plane and is calculated for the wavelengths of radiations at E = 2GeV and ρ = 5.55m. Right: Comparison between theory and measurement of the angular dependence of the photon flux, as obtained by Codling and Madden (1965) [28]. The intensity radiated in each component of polarization for monoenergetic electrons (120MeV; λ = 5000Å) is shown.

Chapter 2   Synchrotron Radiation

Time structure

In section 2.2.2 the length of a pulse of the synchrotron spectrum emitted by a single electron is derived (formula (2.30)). In practice many electrons circulate in compact groups (bunches). The bunch length is typically 50ps to 1ns and the radiation is emitted in short flashes of bunch length time. For spectroscopic experiments the repetition frequency of the filled bunches is an important parameter. In the single bunch mode (only one bunch is circulating) maximum separation between two pulses is obtained. In this case the period of revolution (in the range of ns to µs) determines the repetition frequency.

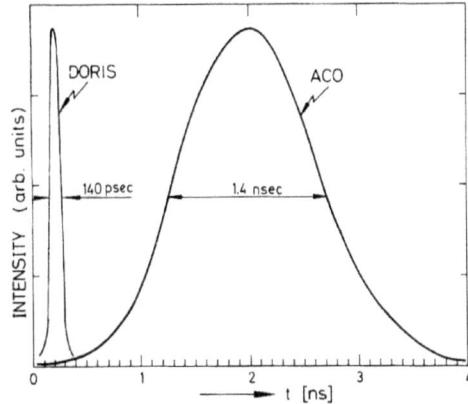

Figure 2.16 (adapted from [26]): Experimentally determined shape of the light pulses for the storage rings DORIS and ACO

Wide spectral range

The spectral range of SR covers energies from the infrared to the X-ray region. Formula (2.32) ($\equiv$ (2.44)) shows that for $E \gg mc^2$ a relativistic particle emits a broad spectrum of frequencies up to γ^3 times ω_0. (ω_0 is the angular frequency of rotation for circular motion.) The frequency distribution of the radiated energy (the frequency spectrum of SR) is given by formula (2.46). The spectral distribution depends only on the particle energy, the critical frequency and the purely mathematical Bessel functions.

Chapter 2   Synchrotron Radiation

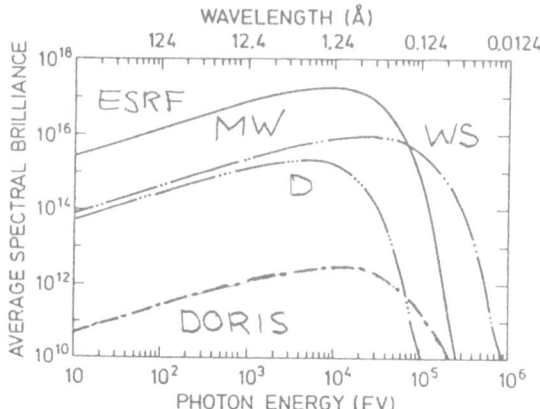

Figure 2.17 (source [24]): Spectral brilliance of dipole radiation emitted from the storage ring DORIS [25] (E = 5 GeV). Additionally the brilliance of the radiation emitted from a dipole magnet (D), a wavelength shifter (WS) and a Multipole Wiggler (MW) is shown.

Polarization

Formula (2.42), which determines the energy radiated per unit frequency interval per unit solid angle already implies the polarization of SR. The two terms in the square brackets correspond to the radiation polarized in the orbital plane and the radiation polarized perpendicular to that plane respectively. From (2.42) formula (2.64) is derived defining the degree of linear polarization P_l of the radiation. According to (2.64) the radiation is purely linear polarized ($P_l = 1$) in the orbital plane ($\theta = 0$). Integration of (2.42) over all frequencies and all angles shows that the intensity polarized parallel to the plane of the orbit is about 7 times larger than the intensity polarized perpendicular to the orbital plane.

Small source size

Section 2.2.2.1 introduced three key parameters for SR: brightness, flux and brilliance. The most important parameter to characterize the synchrotron source is the brilliance (2.63) because it accounts for the size and angular spread of the electron beam (the source size). It is evident that the angular divergence of the emitted radiation is the better the smaller the source size is. This results in a higher brilliance. The lowest limit for the divergence of SR is given by the intrinsic angular width of the radiation emitted by a single electron in ideal circular motion.

2.3 Insertion devices

A synchrotron radiation facility typically consists of a linear accelerator (LINAC), a synchrotron and a storage ring. The electrons are accelerated in the LINAC and the synchrotron up to a final storage energy and then injected into the storage ring. In the storage ring the high energy electrons are stored for several hours while moving in a definite circular orbit. A synchrotron storage ring in principle consists of three different sections. Along the electron orbit magnetic dipoles (so called bending magnets) provide a homogeneous magnetic field to keep the charged particles on a circular trajectory and are arranged to keep them on a closed orbit. At the location of the dipole magnets beam ports are located so that the synchrotron radiation may be utilized. Additionally, between the bending magnets, there are straight beam line sections, where quadrupole- and sextupole magnets focus and stabilize the electron beam. The electron bunch circulating in the storage ring loses energy by emitting SR. To compensate for this loss in energy, Radio Frequency (RF) accelerating cavities are installed on a straight section of the ring. The radiofrequency is selected to be an integer multiple of the electron orbital revolution frequency (the so called harmonic number of the ring). This allows to group the electrons in the ring in bunches of the harmonic number.

In second and third generation synchrotrons additional straight sections have been designed to insert special devices. These insertion devices are other magnetic structures, which for many experiments are more effective sources of SR than the dipole bending magnets. Utilizing these devices it is possible to have a more intense source than produced by bending magnets, to change or extend the shape of the spectral distribution, to change the polarization features or to have an X-ray beam with smaller divergence. In the following only the features of these special insertion devices will be discussed since the properties of SR emitted by bending magnets have been discussed in the previous chapters.

2.3.1 Wavelength shifters

As their name implies, wavelength shifters are used to extend the emitted spectrum to higher photon energies. To maximize the desired effect, they are often constructed using superconducting technology to allow the production of hard X-rays from electrons with moderate energy. A wavelength shifter consists of three or five (superconducting) dipole magnets arranged as a linear array with alternating magnetic field directions. Figure 2.18 shows schematically a three-pole wavelength shifter.

Chapter 2   Synchrotron Radiation

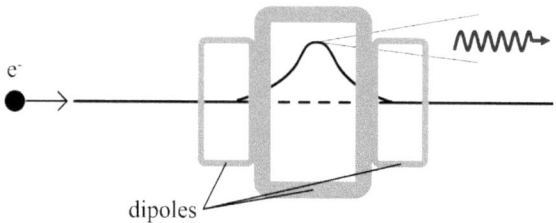

Figure 2.18: Schematic view of a 3-pole wavelength shifter

The central dipole has a higher magnetic field compared to the others and is used as radiation source. The energy of the emitted photons is strongly dependent on the radius of the trajectory and therefore on the local magnetic field. The two side poles are used to compensate the beam deflection caused by the central one (in a five-pole wavelength shifter the three central poles are used as radiation source, while the end poles act as compensators). The electron beam passing through this insertion device is deflected up and down (or left and right) such that no net deflection remains. Therefore this device is neutral on the geometry of the beam path through the storage ring and in principle the deflection can be made as strong as required.

2.3.2 Wigglers

Concerning wiggler-magnets the principle of a wavelength shifter is extended. This insertion device consists of a series of equal dipole magnets with alternating magnet field directions. Like in the wavelength shifter the end poles have to compensate the net deflection to make this device neutral to the geometry of the particle beam path. The main advantage of using many dipole magnets is an increased photon flux. In figure 2.19 a schematic view of a wiggler is shown.

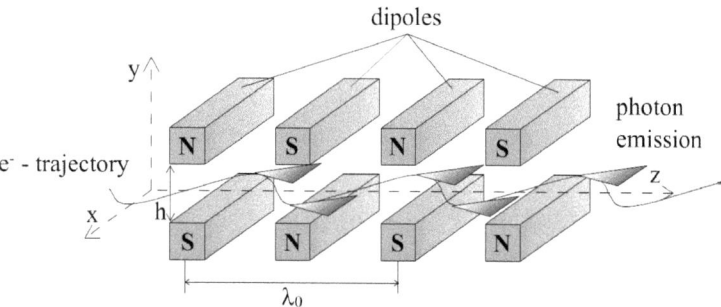

Figure 2.19: Working principle of a multipole wiggler

26

As an electron beam passes through the array of N dipole magnets it is forced to an almost sinusoidal trajectory. Like in a single bending magnet, each of the N magnet poles forces the electron to change its direction producing a fan of synchrotron radiation of critical energy ε_c in the forward direction. The radiation emitted at each dipole occurs as a sequence of short pulses and is incoherently superimposed. The resulting total flux is N-times larger than that from a single pole.

To distinguish between multipole wigglers and undulators, which are similarly constructed, the so called deflection parameter K is defined as

$$K = \frac{eB_0 \lambda_0}{2\pi mc^2} \tag{2.66}$$

where λ_0 is the magnet period length and B_0 the peak magnetic field.

Multipole wigglers typically have values of K>>1, which is accomplished by high magnetic fields and long magnet period lengths λ_0. The maximum deflection angle δ of the electron beam relatively to the z-axis is given by

$$\delta = \frac{K}{\gamma} \tag{2.67}$$

leading to a horizontal opening angle of $2\delta = 2K/\gamma$.

2.3.3 Undulators

Undulators are constructed similarly to multipole wigglers (figure 2.19), but with a deflection parameter K smaller than unity. This is realized using a larger number of dipoles with weaker magnetic fields. The particle deflection in an undulator is small compared to a wiggler and the deflection angle is comparable or less than that of the SR emission cone ($2/\gamma$; see equations (2.16) and (2.19)). The radiation emitted at each turning point of the electron beam is coherently superimposed and strong interference effects influence the spectral and spatial distribution. In the emission spectrum sharp peaks with discrete wavelengths λ appear, given by

$$\lambda = \frac{\lambda_0}{2j\gamma^2}\left(1 + \frac{1}{2}K^2 + \gamma^2\theta^2\right), \qquad j = 1, 2, 3, \ldots \tag{2.68}$$

where θ is the angle between the z-axis and the direction of observation. The spectral intensity of the radiation is proportional N^2, where N is the number of dipoles and the width of the spectral line is proportional to $1/jN$.

Figure 2.20 shows the different properties of synchrotron radiation emitted from bending magnets and different insertion devices.

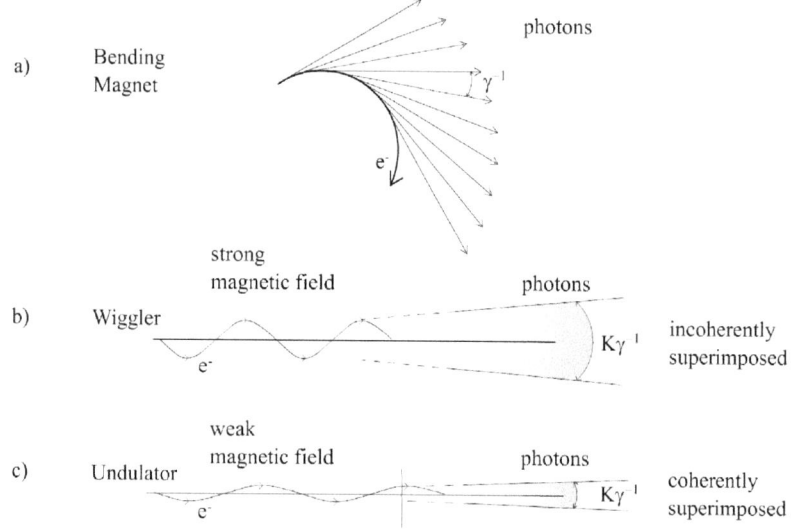

Figure 2.20: Comparison of emitted synchrotron radiation from a) bending magnets, b) wigglers and c) undulators

In figure 2.21 the brilliance of bending magnets and various insertion devices of different synchrotron radiation sources (HASYLAB, BESSY, ESRF) is compared.

2.3.4 Free electron lasers

Recently, with the development of X-ray free electron lasers (XFEL) a new generation of radiation sources is created. These sources are characterized by an extremely high brilliance and a time structure with pulses in the range of femto-seconds. The principle setup of an XFEL consists of a very long undulator. Due to the interaction of the oscillating electrons with their emitted radiation, the electrons are merged into so called microbunches. These microbunches are separated by a distance equal to one wavelength of the radiation. This leads to constructive interference of the emitted radiation and therefore to extreme intense and short light pulses. This effect is called Self-Amplified Spontaneous Emission (SASE) which is the underlying principle of the XFEL. Figure 2.22 shows a comparison of the brilliance of X-ray free electron lasers (TESLA XFEL, TTF VUV-FEL and LCLS) in comparison with undulators at present third generation synchrotron radiation sources.

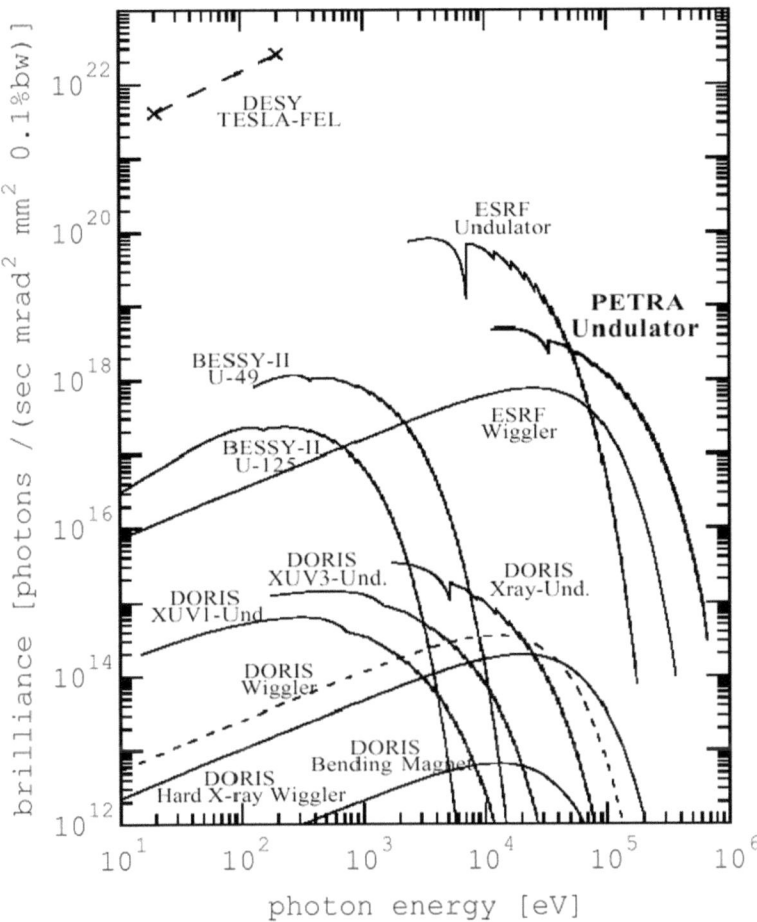

Figure 2.21 (source [29]): Brilliance of bending magnets and different insertion devices at different synchrotron radiation facilities (HASYLAB, BESSY, ESRF)

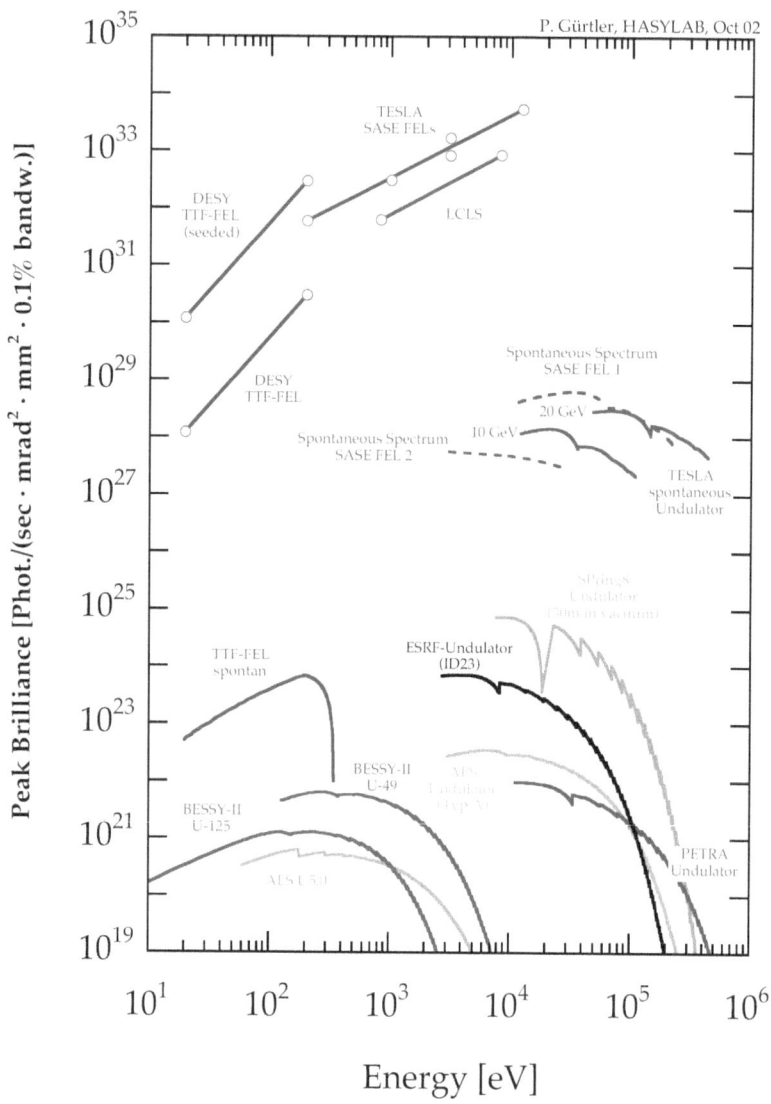

Figure 2.22 (adapted from [30]): Brilliance of X-ray free electron lasers (TESLA XFEL, TTF VUV-FEL and LCLS) in comparison with undulators at present third generation synchrotron radiation sources

Chapter 3

X-Ray Absorption Fine Structure and Total Reflection X-Ray Fluorescence Analysis

3.1 X-Ray absorption and fluorescence

X-Ray photons (energy region: ~100eV - 100keV) traveling through matter are absorbed according to Beer-Lambert's law:

$$I(x) = I_0 e^{-\mu(E)x} \qquad (3.1)$$

Where $\mu(E)$ is the linear attenuation coefficient [cm^{-1}], x [cm] is the thickness of the interfused matter, $I(x)$ is the transmitted and I_0 is the incident X-Ray beam intensity.

Basically three effects are responsible for the loss of intensity in this energy region: the photoeffect, elastic, and inelastic scattering. The linear attenuation coefficient (in the following called simply "absorption coefficient") can therefore be written as a sum of the photo-absorption coefficient τ, the coherent (σ_{coh}) and the incoherent scattering coefficients (σ_{incoh}):

$$\mu(E) = \tau(E) + \sigma_{coh}(E) + \sigma_{incoh}(E) \qquad (3.2)$$

As shown in the right part of figure 3.1 the absorption coefficient μ is a smooth function at most energies. However, if the incident photon reaches energies in the region of the binding energy of a core-level electron it has a certain probability to remove the electron to the continuum (photoeffect). In correspondence of such a binding energy the absorption increases abruptly and an absorption edge is formed. Following an absorption event, the atom is in an excited state, with a so-called core-hole, and a photo-electron. The excited state decays within a few femtoseconds following on of two main mechanisms: the emittance of an Auger-electron or a characteristic fluorescence photon. This is the basis of Auger- and fluorescence spectroscopy.

Chapter 3   X-Ray Absorption Fine Structure & Total Reflection X-Ray Fluorescence Analysis

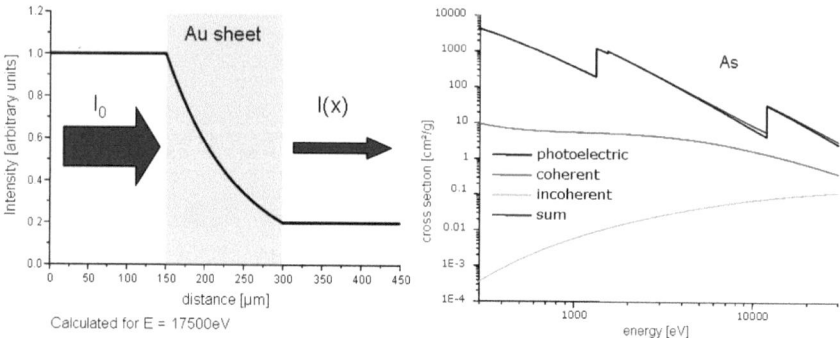

Figure 3.1: Left: Attenuation of X-Ray intensity calculated for photons with 17.5keV energy passing through a Gold-foil of 150μm thickness. Right: Mass attenuation coefficient μ/ρ [cm²/g] for Arsenic calculated for an energy region of 300eV-30keV.

3.2 X-Ray absorption fine structure

3.2.1 Introduction

The fine structure of the absorption coefficient near an absorption edge was already observed in the 1920s when X-ray absorption spectroscopy was first used for structural investigations of matter. The fine structure observed near the edge was referred to as "Kossel structure" for many years as a first explanation for its origin was given by a theory of Kossel [31]. The structure several hundreds eV above the edge was called "Kronig structure" because it was firstly explained theoretically by Kronig [32]. A detailed historical overview of the development of the analysis of the X-ray absorption fine structure can be found in [33, 34].

Today the "Kronig structure" is called Extended X-ray Absorption Fine Structure (EXAFS) [35-40] and refers to the oscillations of the total X-Ray absorption coefficient observed over an energy range of several hundred eV starting ~50-100eV above the absorption edge of the element of interest. X-Ray Absorption Near Edge Structure (XANES) analysis [39-42] on the other hand refers to the fine structures around the absorption edge (up to ~50-100eV above the edge) including as well the pre-edge region (figure 3.2). X-Ray Absorption Fine Structure (XAFS) is a general terminology for both EXAFS and XANES techniques [43].

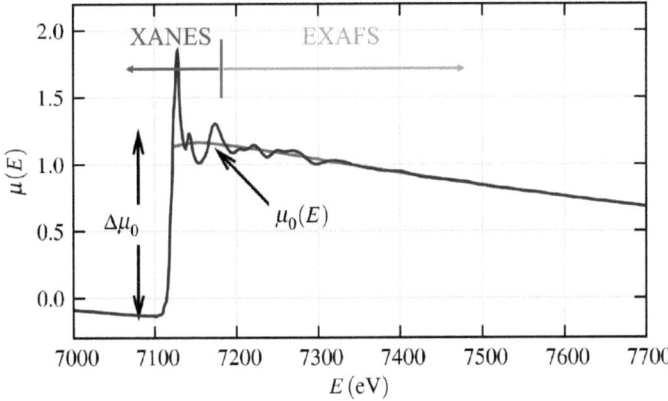

Figure 3.2 (adapted from [44]): XAFS spectra of FeO (blue) showing the XANES and EXAFS regions as well as the edge-step $\Delta\mu_0(E_0)$ and the smooth background function $\mu_0(E)$ (red) representing the absorption of an isolated atom

3.2.2 Theory

While diffraction techniques provide information about the atomic coordinates as a macroscopic average of a long range ordered periodical structure, XAFS gives information on the local atomic structure, i.e. the radial distribution around a particular atomic species and its electronic states. Therefore short range ordered systems can be studied. This capacity derives from the physical origin of XAFS: the interference of outgoing and scattered photoelectron waves (with reference to the excited central atom) modulates the matrix element of the dipole transition. Here exact spherical waves [43] or curved waves [45] are used to describe the propagation of the photoelectrons (instead of the plane wave approximation utilized in earlier models).

The absorption of an X-Ray photon can be interpreted as a transition between two quantum states: the initial state i (photon, core electron, no photo-electron) and the final state f (no photon, core hole, photo-electron). Therefore the absorption coefficient μ(E) can be expressed using Fermi's Golden Rule:

$$\mu(E) \propto \left|\langle i|H|f\rangle\right|^2 \rho(E_f) \qquad (3.3)$$

where H is the interaction Hamiltonian in the matrix element of interaction and $\rho(E_f)$ is the density of final electronic states. In figure 3.3 a schematic representation of the absorption process is shown.

Chapter 3   X-Ray Absorption Fine Structure & Total Reflection X-Ray Fluorescence Analysis

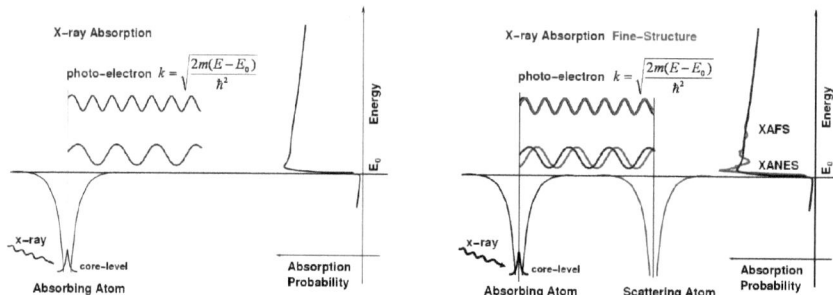

Figure 3.3 (adapted from [44]): X-Ray absorption of a photon with an energy in the region of the binding energy of a core-level electron. The propagation of the photo-electron is shown without (left) and with (right) the presence of a neighboring atom. XAFS occurs because the photo-electron can scatter from a neighbouring atom and returns to the absorbing atom, modulating the amplitude of the photo-electron wave-function at the absorbing atom. This in turn modulates the absorption coefficient µ(E), causing the fine structure.

Absorption only takes place if there is an available state for the photo-electron: i.e. a quantum state at exactly the right energy, and also the right angular momentum state. If a photo-electron with wave number k is created, it propagates away from the atom. This photo-electron can scatter from the electrons of a neighbouring atom and returns to the absorbing atom. The interference between outgoing and scattered photoelectron waves affects the available final states in the matrix element of the dipole transition of the excited atom. It follows that the presence of the photo-electron scattered back from neighbouring atoms alters the absorption coefficient. This is the origin of the fine structure of µ(E).

Regarding the EXAFS regime far from the edge the photoelectrons mean free path is short and the scattering amplitudes are weak. Therefore it is usually adequate to consider single-scattering contributions only, which results in a remarkable simplification of the theory. Close to the edge, in the XANES region, multiple-scattering effects become more important as the mean free path of the photoelectrons is longer and the scattering strengths are greater. Hence the near edge region oscillations are dominated by multiple-scattering resonances of photoelectrons within a cluster around the excited atom. In contrast to that, the information obtained by EXAFS analysis can be interpreted as more local as the interference between the photoelectron wave functions is dominated by contributions from single-scattering and double-scattering paths. Contributions of multiple-scattering beyond double-scattering are often neglected because the spherical wave emitted from the central atom is rapidly damped due to inelastic effects. However, in order to achieve exact structural information beyond nearest neighbor distances it is essential to consider multiple-scattering effects also for EXAFS. Today high order multiple-scattering calculations are performed using *ab initio* calculation software like FEFF [46] which is based on the algorithms developed by Rehr and Albers [47].

In order to analyze the measured X-Ray absorption fine structure the oscillations above the edge have to be extracted. This is done using the so called XAFS function $\chi(E)$ which is defined as:

$$\chi(E) = \frac{\mu(E) - \mu_0(E)}{\mu_0(E)} \tag{3.4}$$

where $\mu(E)$ is the measured absorption coefficient and

$$\mu_0(E) = |\langle i|H|f_0\rangle|^2 \tag{3.5}$$

is a smooth background function representing the absorption of an isolated atom (see figure 3.2). The XAFS function is usually transformed to k-space using:

$$k = \sqrt{\frac{2m_e(E - E_0)}{\hbar^2}} \tag{3.6}$$

Here k is the photo-electron wavenumber, m_e is the electron mass and E_0 is the binding energy.

In figure 3.4 the isolated XAFS function $\chi(k)$ of the EXAFS determined for FeO (figure 3.2) is shown. The oscillations in $\chi(k)$ are composed of different frequencies related to different near-neighbor atom types ("coordination shells") and can be described according to the EXAFS-equation (3.7). Therefore a Fourier transformation can be applied to extract information about the inter-atomic distances and the number of the surrounding atoms (i.e. the radii r_n of the near-neighbor coordination shells and the number of atoms at r_n). The EXAFS equation can be derived from Fermi's Golden Rule using first order perturbation theory and the dipole approximation and is given by (e.g. [37, 40, 44, 48]):

$$\chi(k) = \sum_j \frac{N_j f_j(k) e^{-2k^2 \sigma_j^2} e^{-2R_j/\lambda(k)}}{kR_j^2} \sin[2kR_j + \delta_j(k)] \tag{3.7}$$

Here j is the index of the coordination shell, N_j is the number of neighbouring atoms at a distance R_j with a mean-square-displacement σ_j^2 (i.e. the disorder in the neighbour distance), λ is the mean free path of the photoelectron, and $f_j(k)$ and $\delta_j(k)$ are scattering properties of the atoms neighbouring the excited atom. $\delta_j(k)$ is often referred to as total phase shift function and is the sum of the absorbing atom phase shift and the scattering atom phase shift. This total phase shift is nearly a linear function of k and therefore the peak maxima of the Fourier transform are shifted towards smaller R values. It is necessary to have accurate values for the scattering amplitude $f_j(k)$ and the total phase-shift $\delta_j(k)$ to get the corrected distances and coordination numbers from the EXAFS. Values for $f_j(k)$ and $\delta_j(k)$ can be found in literature (e.g. [49]) or - which is more common nowadays - calculated with different programs like FEFF [46] or GNXAS [50].

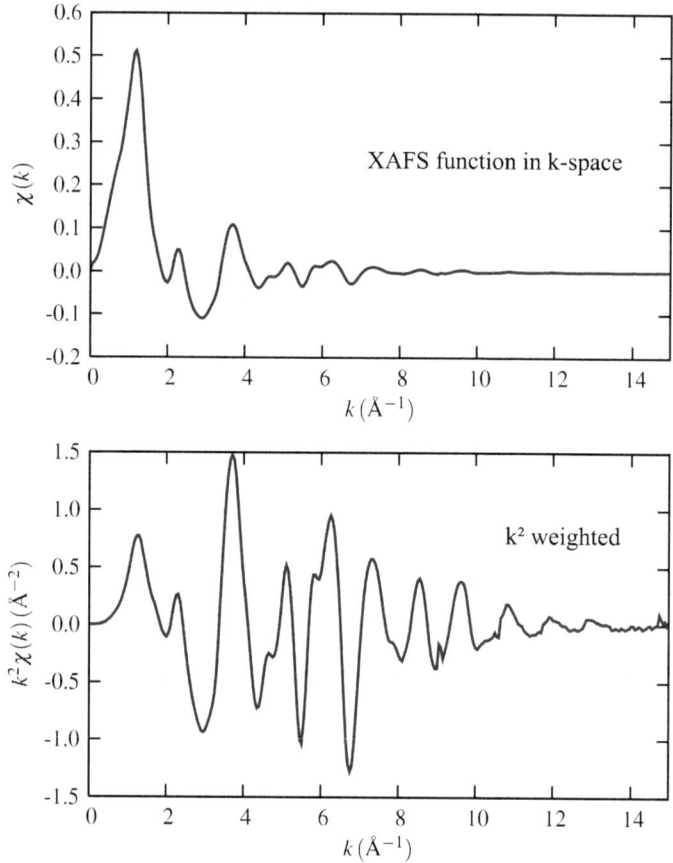

Figure 3.4 (adapted from [44]): Top: XAFS function in k-space extracted from the measured EXAFS of FeO shown in figure 3.2. Bottom: The k-weighted function $k^2\chi(k)$.

The term $\exp(-2k^2\sigma_j^2)$ is an approximation of the Debye-Waller factor which is partly due to thermal effects and causes the atoms to vibrate around their equilibrium position. An additional contribution to σ_j^2 comes from structural disorder. The Debye-Waller factor becomes more important for shorter wavelengths of the photoelectron and cuts off the EXAFS at higher energies ($k \sim 1/\sigma$; typically ~ 10 Å^{-1}). It is therefore relevant in EXAFS but often negligible in XANES. From the $\lambda(k)$ and the R^{-2} term in equation (3.7) it is obvious that

EXAFS is seen to be a local probe and is not able to deliver information from distances much further than a few Angstroms from the central atom. Further it can be seen, that the EXAFS oscillations consist of different frequencies corresponding to the different radii of each coordination shell (which leads to the application of the Fourier analysis).

Figure 3.5 shows the XAFS function $\chi(R)$ derived from the Fourier transformation of the XAFS shown in figure 3.4. Here two main peaks can be clearly identified which correspond to the Fe-O and the Fe-Fe bond distances of the FeO compound. The first peak occurs at 1.6 Å although the Fe-O distance in FeO is known to be 2.14 Å. This is due to the above mentioned phase-shift $\delta_j(k)$ which is typically in the range of 0.5 Å.

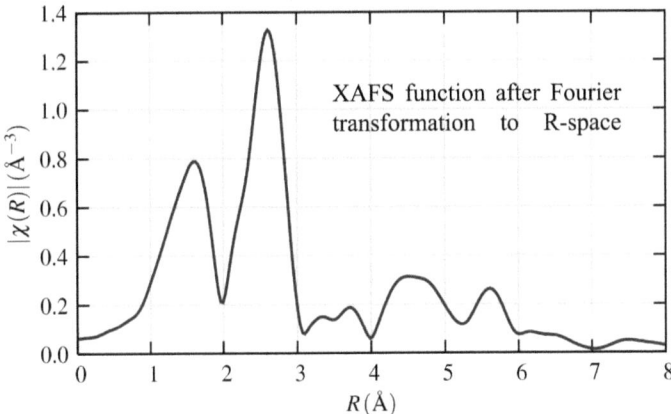

Figure 3.5 (adapted from [44]): The Fourier transformed XAFS $\chi(R)$ for FeO derived from the XAFS function shown in figure 3.4. The data is not phase-shift corrected.

Since the scattering factor is larger in the low energy region, the magnitude of XANES modulation is generally larger than that of EXAFS. In the near-edge region multiple scattering becomes dominant which is sensitive to the bond angle. Furthermore the scattering of low energy electrons is dependent on the shape of the potential and is therefore sensitive to chemical bonds or valence. This implies that the XANES contains additional information on site symmetry and electronic states but is much more complex to interpret. All possible multiple scattering pathways have to be considered to calculate the near edge structures. However, the main difficulty is that the EXAFS equation (3.7) breaks down at low values of k, due to the increase in the mean-free-path and the 1/k term. The theory of XANES requires different physical considerations and a multiple-scattering treatment seems not fully sufficient so far because of atomic, chemical, and many-body corrections [40]. Since a complete theoretical description of XANES is still problematic, it is common to use XANES analysis as a "finger printing" technique (i.e. comparison of measurements of unknown samples with known reference compounds). The best theoretical approach seems to be supposing that XANES essentially measures the projected density of states. This follows the fact that the X-Ray absorption coefficient is equivalent to the golden rule of Fermi, and can be written in terms of the projected photoelectron density of final states. When utilizing this interpretation of XAFS one has to be aware of final-state effects like core-hole effects, lifetime broadening, or energy shifts. However, modern full-potential electronic-structure methods using a local-density-functional approximation (LDA) have some advantages regarding the calculation of XANES [51]: Adapted to the structure of the atomic environment they provide accurate self-consistent calculations of the electronic charge densities and potentials. Furthermore they provide a precise calculation of the Fermi energy level relative to the unoccupied states. On the other hand these full-potential methods are not well applicable to EXAFS because it is very difficult to include effects which are import for calculations beyond 50-100eV of the absorption edge (core-hole potentials, final-state lifetime effects, Debye-Waller factors) [40].

3.2.3 Experimental

A great advantage of X-Ray absorption spectroscopy (XAS) is the selectivity for a particular atomic species in the sample; the energies of various inner shells are distributed over a wide range in energy and one can tune to a particular species by choosing the energy of incident photons. This leads to the demand of an intense and energy-tunable X-Ray source and indicates that the capability of XAS is strongly dependent on the characteristics of the light source. In XAFS a typical magnitude of modulation is of the order of 10^{-2} and the statistical error must be less than 10^{-3}; therefore one needs to collect at least 10^6 photons for the incident and transmitted beam intensities. XAFS measurements require a high photon flux because the intensity is usually reduced by several orders of magnitude after the absorption. Thus the advance of XAS was strongly associated with the application of synchrotron radiation (SR). SR has outstanding advantages as a light source for XAS, namely i) an intense quasi-parallel beam which allows measuring XAFS with high energy resolution; ii) a high radiation purity and a wide spectral range covering the VUV and soft X-Ray region (250eV - 4keV) to the hard X-Ray region (4keV - 30keV), and iii) polarization characteristics with high purity and controllability which can be used for polarization dependent experiments. Furthermore, due to the high flux delivered by synchrotron radiation sources the data collection time is decreased drastically and high energy resolution measurements are feasible. Fixed exit double-crystal monochromators composed of perfect crystals like Si(111) or Si(311) deliver energy resolutions ΔE/E in the range of 10^{-4} which is generally sufficient for XAFS analysis [37].

XAFS measurements are commonly performed in transmission, fluorescence, or non-radiative (Auger- or secondary electron) detection modes. The transmission experiment represents the classical XAS setup where a monochromatized X-Ray beam passes through the sample. The parameter of interest, the absorption coefficient μ(E), is obtained as a function of the photon energy E from the difference between incident (I_0) and transmitted (I) beam intensities according to Beer-Lambert's law (see also equation (3.1)):

$$\mu(E) = -\ln(I/I_0) \qquad (3.8)$$

For uniform samples the absorption coefficient is the sum of the absorption of the atom of interest $\mu_i(E)$ - which shows a fine-structure - and a contribution of the other atoms which is assumed to be constant in the energy region analyzed. The intensities I_0 and I are measured using linear detectors such as ionization chambers. Transmission experiments are preferred for higher concentrated samples where the element of interest is a major component. Here the sample thickness has to be well chosen to get a considerable signal for the transmitted intensity.

X-Ray fluorescence or non-radiative detection is usually performed for thick or (highly) diluted samples (samples with concentrations in the ppm range or even lower) because of its higher sensitivity. The fluorescence (or Auger) yield is proportional to the absorption coefficient:

$$\mu(E) \propto I_f / I_0 \qquad (3.9)$$

where I_f is the monitored intensity of a fluorescence line (or electron emission) associated with the absorption process. This relation is derived from the formula for the intensity $I_f(E)$ of the fluorescence line of the element of interest as a function of the incident photons' energy E (assuming monoenergetic excitation) which is accepted by an energy dispersive detector with solid angle $\Omega/4\pi$:

$$I_f(E) = I_0 \mu(E) \varepsilon_f \frac{1}{\sin\theta} \int \frac{d\Omega}{4\pi} \int_{x=0}^{x=t} e^{-A(E)x} dx \qquad (3.10)$$

with $A(E) = (\mu(E)/\sin\theta) + (\mu(E_f)/\sin\varphi)$

Here E is the energy of the incident and E_f the energy of the fluorescent photon; ε_f is the fluorescence yield; t is the sample thickness; θ is the angle between sample surface and incident beam, and φ is the angle between sample surface and detector; (secondary excitation and detector efficiency are neglected as well as absorption of the fluorescence radiation between sample and detector). Assuming an infinitely thick (x=0 to x=∞), uniform sample the integration over x in equation (3.10) becomes 1/A(E). On the other hand, if the sample is very thin (x→0; thin film approximation) the integration term turns to 1. Hence the intensity becomes proportional to 1/sinθ but is no longer dependent on φ. Here it has to be noted that a grazing incidence (GI) angle φ (below the critical angle of total reflection, see below) was assumed. The thin-film approximation is therefore valid for grazing incidence geometry which in turn represents a surface (thin-film) sensitive technique. Such a surface sensitive detection scheme is required to apply XAFS analysis to surfaces and buried interfaces because absorption experiments in general and fluorescence detection using a conventional geometry are not surface sensitive in the hard X-Ray region (>4keV) due the large penetration depth of the incident photons. Due to the special glancing incidence geometry an effect occurs which is the basis of a powerful technique for the multi-element analysis of traces: the total reflection of X-Rays.

3.3 Total reflection X-Ray fluorescence analysis

3.3.1 Introduction

The phenomenon of total reflection of X-rays has been experimentally discovered in the 1920s by Compton ([52] reprinted in [53]) who observed that the reflectivity of a flat target was enlarged below an angle of ~0.1°. However it took almost 50 years until this effect was utilized for XRF analysis by Yoneda and Horiuchi [54] in 1971 by applying a small sample amount on top of a flat reflector. Subsequently this technique was refined by Wobrauschek in his PhD thesis [55] followed by several authors [56-59] and called "Total Reflection X-Ray Fluorescence Analysis" (TXRF).

Shortly after TXRF was introduced as spectrometric technique it was pointed out by Andersen et. al [60] in 1976 that X-ray standing waves originating from the interference of incident and reflected beam extend above the surface of a crystal. This effect was used by Cowan et al. [61] (1980) to locate the position of an adsorbed atom layer on a perfect single-crystal surface. Later Bedzyk et al. [62] (1989) demonstrated how an X-ray standing wave with a variable period can be generated by total external reflection of a monochromatic beam from a reflective surface. This technique was used to locate a layer of heavy atoms embedded in a low-Z thin film deposited on the reflectors surface.

In 1954 Parrat already used the angle dependence of the intensity of the (total) reflected X-Ray beam to investigate surfaces [63]. But the angular dependence of the fluorescence intensity for angles in the range of the critical angle of total reflection was first observed in 1983 by Becker et al. [64]. The authors also presented first results of the inverse geometry – the grazing exit arrangement – which will be discussed later. Today this effect (the angular dependence of the fluorescence intensity) is used to investigate surface impurities, thin surface layers and multilayer structures and is called "Grazing Incidence" or "Glancing Incidence" (GI) XRF [65].

The use of Synchrotron Radiation (SR) is highly beneficial for angle-dependent XRF and for TXRF in general as its properties like high intensity, linear polarization, small source size, and natural collimation make it an ideal radiation source. First experiments have been published 1986 by Iida et al. [66] followed by results of several other authors [67-69]. The sensitivity of Synchrotron Radiation induced TXRF (SR-TXRF) is several orders of magnitudes higher than the one of TXRF using conventional sources like X-Ray tubes [68, 70-76].

Review articles providing more details about development and applications of TXRF and SR-TXRF have been published recently by Wobrauschek [75] (2007) and Streli et al. [76] (2008) respectively.

Today TXRF is mainly used for surface contamination analysis in semiconductor industry and for chemical trace analysis as it offers detection limits in the pg range for excitation with X-Ray tubes and even in the fg region if synchrotron radiation is used. Small quantities of solutions or suspensions are placed on sample carriers which show sufficient optical flatness (e.g. Silicon wafers, Quartz or Plexiglas reflectors). During the drying process the residue on the surface of the reflector forms a very thin sample. Subsequently the sample is excited under glancing angle (below the critical angle of total external reflection) and the emitted fluorescence radiation is collected by an energy dispersive detector. The typical experimental arrangement is shown in figure 3.6.

Chapter 3   X-Ray Absorption Fine Structure & Total Reflection X-Ray Fluorescence Analysis

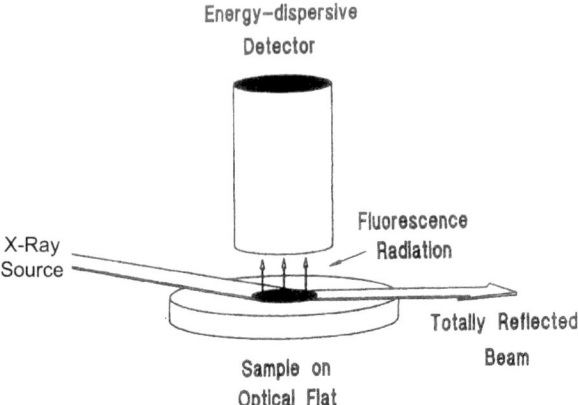

Figure 3.6 (adapted from [72]): Scheme of a typical experimental setup for TXRF analysis

In the following the fundamental advantages of TXRF analysis are presented:
- Due to the total reflection of the incident photons only a very small part of the primary beam penetrates into the sample carrier. This leads to a drastically reduced spectral background contribution which originates from scattering on the substrate.
- As the incident beam is totally reflected on the sample carrier the sample is excited by both the incident and the reflected beam resulting in doubled fluorescence intensity.
- The extreme grazing incidence geometry allows placing the detector very close to the samples surface. This results in a large solid angle for the detection of the fluorescence radiation.

As a consequence the sensitivity in TXRF analysis is very high and the limits of detection are improved by several orders of magnitude when compared to conventional XRF analysis. Furthermore the angular dependence of the fluorescence radiation for angles in the range of the critical angle of total reflection can be utilized to investigate surface impurities, thin near-surface layers, and even molecules adsorbed on flat surfaces.

3.3.2 Theory

The theoretical basis of TXRF can be derived analogous to the theory of light optics. In the following only the outline of the calculations will be presented, details can be found in e.g. [77-79]. Starting point is the complex refraction index n which can be deduced from quantum mechanics:

$$n = 1 - \delta - i\beta \qquad (3.11)$$

Here both scattering and absorption are taken into account by introducing the factors δ and β associated with dispersion and absorption respectively. This correlation becomes apparent when expressing the refractive index n as a function of the scattering factors f_1 and f_2:

$$n = 1 - N_A \frac{r_0 \lambda^2}{2\pi} \frac{\rho}{A} (f_1 + if_2) \qquad (3.12)$$

with $r_0 = e^2/m_0 c^2$ as the classical electron radius.

Here N_A is Avogadro's number, e and m_0 are the electric charge and the rest mass of the electron respectively, c is the velocity of light, λ is the wavelength of the incident radiation, ρ is the density of the medium and A is the atomic mass. The scattering factors f_1 and f_2 are functions of the atomic number and the energy of the incident radiation and tabulated values can be found e.g. in [80]. In the energy region of X-Rays the real part of the complex refraction index (1-δ) is slightly smaller than unity with values of δ in the range of 10^{-5} to 10^{-6}. The imaginary part β is usually smaller than δ, proportional to the photo-absorption coefficient τ and therefore related to absorption (see equation 3.2):

$$\beta = N_A \frac{r_0 \lambda^2}{2\pi} \frac{\rho}{A} f_2 = \frac{\lambda \rho}{4\pi} \tau \qquad (3.13)$$

In the next step one has to consider the interaction of an electromagnetic wave hitting the interface between vacuum and a medium with refractive index n as shown in figure 3.7. The following calculations are based on classical dispersion theory and assume a perfectly flat interface between the two media. Although a real reflector surface shows a certain roughness on a microscopic scale experimental results have shown good agreement with theory.

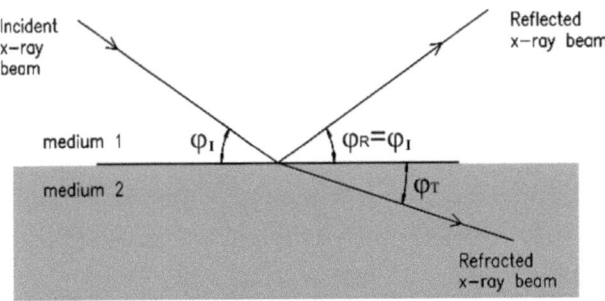

Figure 3.7: Sketch of optical paths of incident, reflected and transmitted beams at the interface of two media. The refraction index of medium 1 (usually vacuum or air) is larger that that of medium 2 (the reflector material).

According to the law of Snellius the refractive angle φ_T is determined by:

$$\frac{n_{medium\,1}}{n_{medium\,2}} = \frac{\cos \varphi_T}{\cos \varphi_I} \qquad (3.14)$$

For X-Rays medium 2 (the reflector material) is optical thinner than vacuum (or air) therefore the refracted beam is deflected towards the boundary (which is contrary to usual light optics). Using the usual expressions of light optics for the electromagnetic waves ($\mathbf{E}=\mathbf{E}_0 \exp(i\mathbf{k}\mathbf{r}-i\omega t)$) of incoming (index I), reflected (index R) and transmitted (index T) beams and utilizing Fresnel's formulas one obtains the ratios of the corresponding amplitudes:

$$\frac{E_T}{E_I} = \frac{2\sin\varphi_I}{\sin\varphi_I + \sqrt{\sin^2\varphi_I - 2\delta - 2i\beta}} \qquad (3.15)$$

$$\frac{E_R}{E_I} = \frac{\sin\varphi_I - \sqrt{\sin^2\varphi_I - 2\delta - 2i\beta}}{\sin\varphi_I + \sqrt{\sin^2\varphi_I - 2\delta - 2i\beta}} \qquad (3.16)$$

As the angles where total reflection occurs are in the range of several mrad the sine functions in equation (3.15) can be replaced by their arguments as a good approximation. For exact calculations the polarization state of the primary radiation and the propagation of the so called inhomogeneous wave have to be considered. However these effects can be neglected for energies higher than that of the soft x-ray region [79]. Using equations (3.15) and (3.16) the reflection (R) and transmission (T) coefficients as well as the refraction angle φ_T can be calculated:

$$R = \left|\frac{E_R}{E_I}\right|^2 = \frac{4X^2(Y-X)^2 + Z^2}{4X^2(Y+X)^2 + Z^2} \qquad (3.17)$$

$$T = \left|\frac{E_T}{E_I}\right|^2 = \frac{16X^3 Y}{4X^2(Y+X)^2 + Z^2} \qquad (3.18)$$

$$\varphi_T = \sqrt{\delta}\sqrt{\sqrt{\left[(Y^2-1)^2 + Z\right]} + (Y^2-1)} \qquad (3.19)$$

with $X = \dfrac{\varphi_T}{\varphi_{crit}}$, $Y = \dfrac{\varphi_I}{\varphi_{crit}}$ and $Z = \dfrac{\beta}{\delta}$ where φ_{crit} is the critical angle of total reflection.

Due to the conservation of energy the reflection and transmission coefficients meet the condition $R + T = 1$. From equation (3.19) it can be seen that the refraction angle is very small for small values of φ_I but does not disappear for $\varphi_I = 0$. The critical angle of total reflection φ_{crit} (total reflection occurs if $\varphi_I < \varphi_{crit}$) can be written as [72]:

$$\varphi_{crit} = \sqrt{2\delta} \qquad (3.20)$$

If R, T or the refraction angle are calculated as a function of the incident angle, the critical angle of total reflection indicates the inflection point of the function (figure 3.8). As δ is inverse proportional to the incident energy (equations (3.11) and (3.12)) it can be seen from equation (3.20) that the critical angle changes with the energy of the incident radiation and depends on the reflector material. The energy dependence of φ_{crit} has to be considered when performing an energy scan of the sample for XAFS analysis.

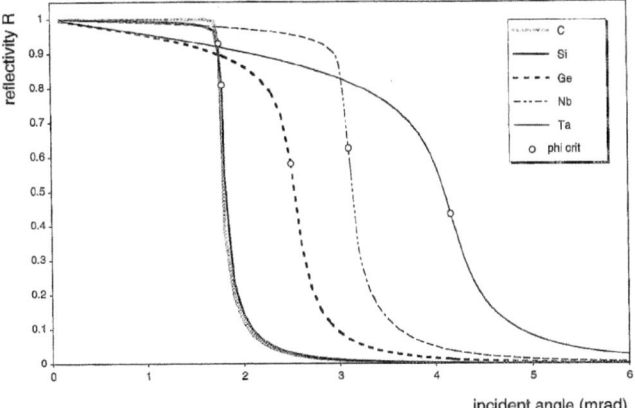

Figure 3.8 (source [74]): Reflectivity calculated for different reflector materials as a function of incident angle. The values of the critical angles (phi crit) are indicated and have been calculated to 1.75, 1.8, 2.5, 3.1 and 4.2mrad for C, Si, Ge, Nb, and Ta respectively.

The penetration depth z_P is defined as the distance measured normal to the samples surface where the intensity of the penetrating (in TXRF the refracted) beam is reduced by a factor of e (see equation (3.1)). For TXRF z_P is therefore direct proportional to the refraction angle (using equation (3.1) and again $\sin \varphi_T \approx \varphi_T$):

$$z_P = \frac{1}{\mu(E)} \varphi_T \qquad (3.21)$$

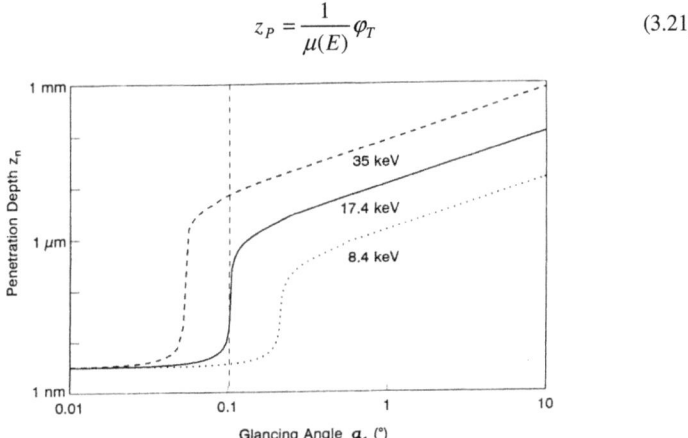

Figure 3.9 (source [72]): Penetration depth of X-Rays in Silicon as a function of the incident angle calculated for three different energies. The dashed vertical line indicates the critical angle for Silicon at 17.4keV.

An important point TXRF analysis is the interference of incident and reflected beam above the reflectors surface. The superposition of the plane electromagnetic waves results in a variation of the intensity depending on the distance above the surface described as standing wave field. The fundamentals are shown in figure 3.10. The length D is the distance between to maxima (or minima) of the standing wave with wavelength λ:

$$D = \frac{\lambda}{2\varphi_I} \qquad (3.22)$$

Typical values for D are in the range of 10-100nm. The intensity of the standing wave is a function of the incident angle and the height z above the reflectors surface (figure 3.11 left):

$$I(\varphi_I, z) = I_0 \left[1 + R(\varphi_I) + 2\sqrt{R(\varphi_I)} \cos\left(\Phi(\varphi_I) - 2\pi \frac{z}{D(\varphi_I)} \right) \right] \qquad (3.23)$$

Here Φ is a phase factor introduced by Bedzyk et al. [62] for the case of a neglected absorption ($\beta=0$):

$$\cos \Phi = 2Y^2 - 1 \qquad (3.24)$$

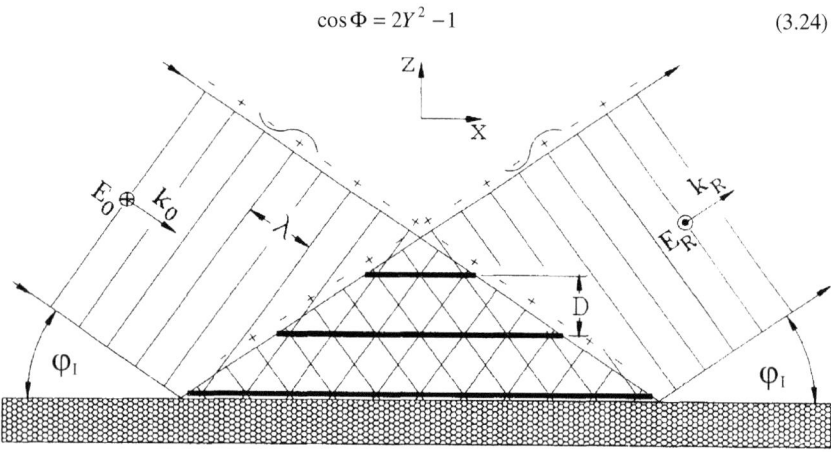

Figure 3.10 (adapted from [62]): The interference zone (standing wave field) between the incident (E_0) and the reflected (E_R) plane waves with wavelengths λ showing nodes and antinodes with a period D.

The intensity distribution of the X-ray fluorescence of the (e.g. silicon) substrate (figure 3.11 right) is the product of the intensity above the surface (figure 3.11 left) and the penetration depth of the radiation. At an angle smaller than the critical angle of total reflection the intensity of the fluorescence radiation is close to zero. At higher angles the intensity increases due to the larger penetration depth of the incident radiation.

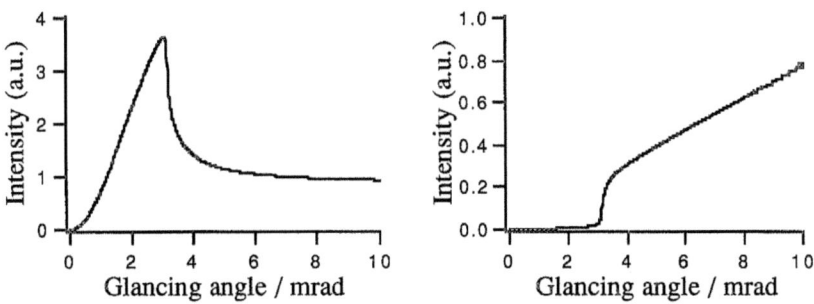

Figure 3.11 (source [81]): Left: Calculated X-Ray intensities above a silicon surface as a function of incident angle. Right: Silicon fluorescence intensities emitted from the substrate as a function of incident angle. Both functions were calculated for an incident radiation with 10keV energy.

Due to this standing wave field the investigation of granular residues on the surface of reflectors with TXRF can be problematic if the thickness of the particles is in the range of the period D. For this case particles with different thickness could give different fluorescence intensities. For samples thicker than D many intensity maxima and minima of the standing wave occur within the sample and will be averaged out (figure 3.12).

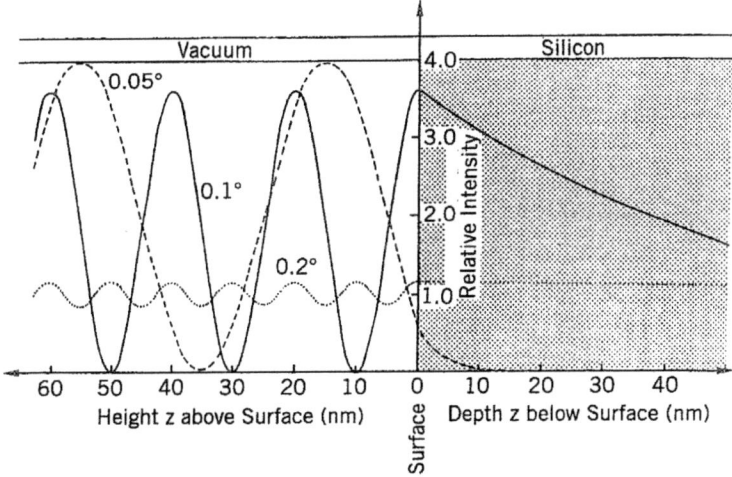

Figure 3.12 (source [72]): X-Ray intensities above (within the standing wave field) and below a thick Si-flat calculated for different angles of incidence. It can be seen that the dependence of the distance D between nodes and antinodes is a function of the incident angle. Inside the medium, the intensity decreases as a function of the refraction angle.

As the intensity of the primary field changes with the angle of incidence it is possible to determine the location of the excited atom which emits the fluorescence radiation by measuring the fluorescence as a function of the incident angle (figure 3.13). The shape of the measured curve allows to distinguish between three sample types, namely a residue on the reflectors surface, an impurity in the surface (buried layer), and a thin layer on the surface.

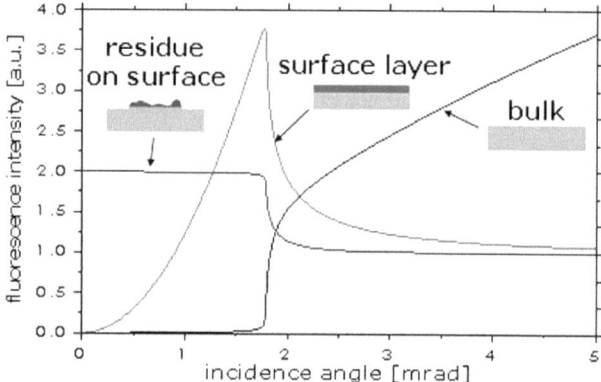

Figure 3.13 (as introduced by [82]): Characteristic shapes due the angular dependence of the fluorescence radiation for three different cases of atomic locations

TXRF spectra are characterized by an excellent peak to background ratio. The background generated by the reflector can be derived from the transmitted part of the primary radiation. Both scattered and fluorescence signal originating from the substrate have a strong dependence on the incident angle. The intensity of the scattered radiation is given by:

$$I_{scatter}(\varphi_I) \propto \varphi_I T(\varphi_I) \frac{d\sigma}{d\Omega} \frac{1}{\mu(E)} \quad (3.25)$$

The first term in equation (3.24) occurs due to a geometry factor proportional to $\sin(\varphi_I)$ ($\approx \varphi_I$). This factor has to be used in TXRF because the area of the reflector which is seen by the detector is generally smaller than the area illuminated by the primary radiation [83]. The transmission coefficient T takes into account that only primary photons which are not totally reflected are able to penetrate in to the substrate. Considering the phenomena of elastic and inelastic scattering, the differential scattering coefficient $d\sigma/d\Omega$ which is a function of the scattering angle (~90° for TXRF) has to be used [84]. Finally the last term considers the attenuation of the refracted beam propagating through the substrate until interaction occurs. The absorption of the radiation on its way from the point of interaction to the reflectors surface toward the detector can usually be neglected.

The geometry factor causes the increase of the spectral background for incident angles larger than the critical angle. Below the critical angle – for total reflection condition - the transmission coefficient T is almost 0 which results in a sudden reduction of the spectral background.

3.3.3 Experimental

3.3.3.1 Synchrotron radiation

Various X-Ray sources can be used for TXRF analysis showing a multiplicity of properties. In this work Synchrotron radiation (SR) was used as excitation source exclusively. Therefore this chapter will focus only on SR induced TXRF (SR-TXRF). To demonstrate the advantages of SR as excitation source for TXRF it is comprehensive to show its influence on the detection limits. The limit of detection (LD) for X-Ray fluorescence analysis is defined as follows:

$$LD = \frac{3 \cdot \sqrt{N_B}}{N_N} \cdot m_{sample} \quad \text{or} \quad LD = \frac{3 \cdot \sqrt{I_B}}{S} \cdot \frac{1}{\sqrt{t}} \qquad (3.26)$$

where N_N and N_B are the net and background counts respectively, m_{sample} is the sample mass, I_B is the background intensity, S is the sensitivity (equal to the net intensity divided by the sample mass) and t is the measuring time.

It is obvious from equation 1 that there are three possibilities to improve the detection limits: i) increasing the sensitivity S (I_N/m), ii) reducing the background or iii) increasing the measuring time, which is of course limited for practical reasons.

The net intensity I_N is – of course – proportional to the intensity of the exciting radiation. Using Synchrotron radiation with its high flux is therefore advantageous in comparison to the application of X-ray tubes. A further increase in sensitivity can be attained by using a tunable excitation source which enables to adjust the exciting energy just above the absorption edge of the element of interest. This results in optimized excitation conditions for this element.

Apart from using total reflection geometry to reduce the spectral background (and doubling the fluorescence intensity), another possibility to reduce the background is to decrease the scatter contributions from the sample itself by utilizing linearly polarized primary radiation [85, 86]. Due to the anisotropic emission characteristics of the scattered radiation based on the classical dipole oscillator emission it is advantageous to place a detector in such a position that only the isotropic emission of the fluorescence signal is detected. Hence the combination of TXRF with polarized radiation leads to a lower background. Further, the use of monochromatic primary radiation improves the background conditions because only photons of one specific energy are scattered.

All these improvements can be accomplished when using Synchrotron radiation which offers new possibilities for improving the power of TXRF. The intense beam with a continuous spectral distribution from photon energies in the infrared region to high energy photons as well as the linear polarization in the orbit plane and its natural collimation are features best suited for excitation in total reflection geometry.

3.3.3.2 Quantification

One of the inherent advantages of TXRF is the fact that the sample forms a thin film on the sample reflector thus no matrix effects have to be considered and the so-called thin film approximation is applicable (both absorption and enhancement effects can be neglected). The intensity of the fluorescent radiation, for e.g. the Kα line of an element i with concentration c_i in the sample (with mass m), is then given by:

$$I(E^i_{K\alpha}) = \int_{E=E^i_{abs}}^{E_{max}} I0(E) G \frac{m}{\sin\varphi} \sigma^i_{K\alpha}(E) c_i f(E^i_{K\alpha}) \varepsilon(E^i_{K\alpha}) dE \qquad (3.27)$$

Here E_{abs} is the energy of the absorption edge of element i, E_{max} is the maximum energy of the excitation spectrum, $E^i_{K\alpha}$ is the energy of the Kα line of element i, $I_0(E)$ is the spectral distribution of the exciting radiation, G is the geometry factor, $\sigma^i_{K\alpha}$ is the fluorescence cross section for the K-shell of element i, $f(E^i_{K\alpha})$ is the absorption factor for the fluorescence radiation between sample and detector, and $\varepsilon(E^i_{K\alpha})$ is the relative detector efficiency for the energy $E^i_{K\alpha}$. It is assumed that the sample is always completely irradiated by the primary radiation.

In this special case, the relation between concentration and fluorescence intensity is linear. Here the sensitivity S [counts/(second)(sample mass)] of the element i can be defined as:

$$S_i = \frac{I_i}{m \cdot c_i} \qquad (3.28)$$

S_i depends only on fundamental parameters and the measuring conditions, which usually can be assumed to be constant. If an element St is used as internal standard the relative sensitivity for element i can be defined as:

$$S^i_{rel} = \frac{f(E^i_{K\alpha}) \varepsilon(E^i_{K\alpha}) \int_{E=E^i_{abs}}^{E_{max}} I_0(E) \sigma^i_{K\alpha}(E) dE}{f(E^{St}_{K\alpha}) \varepsilon(E^{St}_{K\alpha}) \int_{E=E^{St}_{abs}}^{E_{max}} I_0(E) \sigma^{St}_{K\alpha}(E) dE} \qquad (3.29)$$

The relative sensitivity can be determined experimentally with artificially prepared standards or calculated theoretically. The determination of the concentration c_i of the element i in an unknown sample spiked with the same internal standard element is now simple:

$$c_i = \frac{I_i}{I_{St}} \frac{1}{S_{rel}^i} c_{St} \qquad (3.30)$$

Equation (3.29) shows the linear correlation between the intensity I_i and the concentration c_i which enables a simple quantification because geometric and volumetric errors can be canceled.

3.3.3.3 Grazing Exit Geometry

Grazing incidence (GI) XRF uses the angle dependent wave field intensity in order to characterize the structure of layered materials and the composition gradient of materials that are inhomogeneous along the direction perpendicular to the surface. The interference between incident and reflected beam causes in case of microcrystalline samples an intensity increase of the fluorescence signal by a factor (1+R) where R is the reflectivity numerically close to 1. The additional effect due to the penetration depth in the nm region is a low background. The inverse GI-XRF with the incident beam perpendicular to the reflector surface and collection of the fluorescence under grazing angles can also be applied [64, 72, 87-90]. This mode of analysis was named grazing exit X-Ray fluorescence (GE-XRF) and is theoretically based on the reciprocity theorem [64]. The interference in this case is not between primary and reflected beam but among the superposition interference of the fluorescent waves emitted from the sample and observed under the critical angle of total reflection. This geometry is - according to the reciprocity theorem - equivalent to the TXRF setup with the drawback of a lower sensitivity due to the much smaller solid angle seen by the detector. Furthermore absorption of the fluorescence radiation on its way out of the sample due to the sample itself becomes more severe because the path length through the sample is significantly increased. An advantage of the GE setup is the possibility to use a focused micro-beam of (synchrotron) radiation for excitation. This allows a spatially resolved investigation of the sample even in combination with XAFS analysis [91].

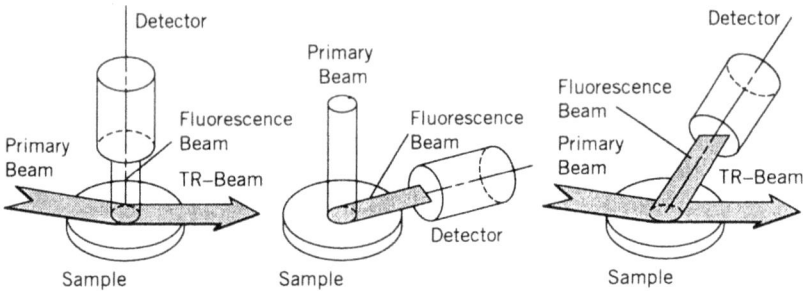

Figure 3.14 (source [72]): Three possible arrangements for TXRF analysis: a) grazing incidence of the primary beam (GI-setup); b) normal incidence of the primary beam and detection of the fluorescence radiation under glancing angle (GE-setup); c) grazing incidence and exit of both beams. (TR= Total reflection)

3.4 Summary

Total reflection X-Ray Fluorescence (TXRF) analysis has proven to be a powerful tool in the analysis of trace elements. It allows a nondestructive, multi element analysis of trace amounts (ng, pg, fg) of many elements and only small sample volumes (µl) are necessary. If the angle of incidence is varied around the critical angle of total reflection (grazing incidence setup) the angle dependence of the fluorescence intensity allows a differentiation between the type of investigated sample, namely a residue on the reflectors surface, a bulk sample, and layers on or in a substrate. The latter enables to determine depth profiles of the concentration of a specific element in the substrate.

Synchrotron Radiation (SR) with its unique properties is an ideal source for TXRF analysis improving the detection limits as well as the performance of angle dependent measurements. SR-TXRF offers sensitivities as high as 34000 cps/ng [92] and detection limits in the fg range for transition metals with a multilayer monochromator and a bending magnet beamline [71, 93-98]. Furthermore, if a crystal monochromator is used instead of a multilayer the technique can be extended to X-ray absorption spectroscopy (XAS) to gain chemical information on a specific element of interest [41, 95, 99-101]. With this modified set-up there is a flux reduction of about two orders of magnitude, but it is still sufficient for X-ray Absorption Near Edge Structure (XANES) analysis at ppb level [95, 99, 102, 103].

This allows an extension of XAS to the trace element level in droplet samples where only small amounts are available [95, 99] and even in the low energy range but using a plain grating monochromator [104]. However, due to the longer path length of the incident beam in the droplet for TXRF geometry, its absorption in the sample cannot be ignored for larger amounts of concentrated samples (see chapter 7). Self-absorption effects have been studied by several authors [105, 106] but their the investigations did not consider different droplet sample geometries. It was suggested that a normal incidence-grazing-exit geometry would not suffer from self-absorption effects in XAFS analysis due to the minimized path length of the incident beam through the sample.

The Grazing Exit (GE) geometry is - according to the reciprocity theorem - equivalent to the TXRF setup with the drawback of a lower s sensitivity [107]. The sensitivity is reduced due to the much smaller solid angle seen by the detector and the absorption of the fluorescence radiation emerging from the sample.

Chapter 4

Silicon Wafer Surface Analysis of Fe contaminations

4.1 Introduction

Iron contaminations on Si wafer surface are known to be a serious limiting factor to yield and reliability of complementary metal oxide semiconductor (CMOS) based integrated circuits [108]. Total reflection X-ray fluorescence (TXRF) is a wide spread analytical technique for the monitoring of surface contamination on non patterned wafers in the semiconductor industry [75, 109]. In laboratory based instruments it offers detection limits down to the 5E9 at/cm2 [110]. When higher sensitivity is requested, monitoring is typically carried out by vapor phase decomposition (VPD) of the native oxide layer and analysis by means of laboratory based TXRF and/or inductively coupled plasma mass spectrometry (ICP-MS) on the expense of the loss of the information relative to the location and distribution of the contamination on the wafer [75, 109, 111-113]. For tracing the source of the contamination not only the distribution of the contamination is very valuable but also additional information on the chemical state of the element can be necessary. Understanding the chemical state is important to gain information on the source of contamination: if more metallic in nature the contamination could be particles from wear or shearing of moving parts; if the iron is an oxide, corrosion maybe taking place. Other species may indicate unexpected chemical reactions taking place.

Synchrotron radiation induced TXRF (SR-TXRF) is a microanalytical technique which offers sensitivities as high as 34000 cps/ng and detection limits in the fg range for transition metals with a multilayer monochromator and a bending magnet beamline [92, 93, 96-98]. With a crystal monochromator the technique can be coupled to X-ray Absorption Spectroscopy (XAS) to gain information on the chemical environment of the specific elements of interest. With this modified set-up there is a flux reduction of about two orders of magnitude, but it is still sufficient for X-ray Absorption Near Edge Structure (XANES) analysis at ppb level [95, 99, 102, 103].

The presented study was motivated by the question whether the known low concentrations of Si-wafer contaminations are sufficient for a SR-TXRF analysis using a Si(111) monochromator and including a XANES analysis within a reasonable time frame. Therefore one aim of the study was to perform the measurements on the points of interest within 48 hours.

4.2 Experimental

A wafer sample from IBM laboratories, showing surface contaminations in the 4E12 at/cm² range for Fe has been investigated (see section 4.3 and figure 4.2). SR-TXRF XANES measurements were performed at the bending magnet beamline L at HASYLAB using a modified micro-XRF setup. This modified setup allowed TXRF measurements with scanning capability. The Wafer with 200 mm diameter was mounted vertically on a custom made holder on a X,Y,Z,θ-stage in front of the side-looking detector to allow area scans and scans over the angle of incidence. The distance between sample and detector was set to 1 mm. A laminar flow hood is present on the set-up at HASYLAB to prevent contamination during the measurement. The Si(111) double crystal monochromator was used and a Silicon Drift Detector (SDD) with 50 mm² active area (VORTEX 50 mm², Radiant Detector Technologies) [114, 115] for the detection of the fluorescence radiation. The beam dimensions were adjusted to 1600 x 400 µm (vertical x horizontal). All measurements were performed in air.

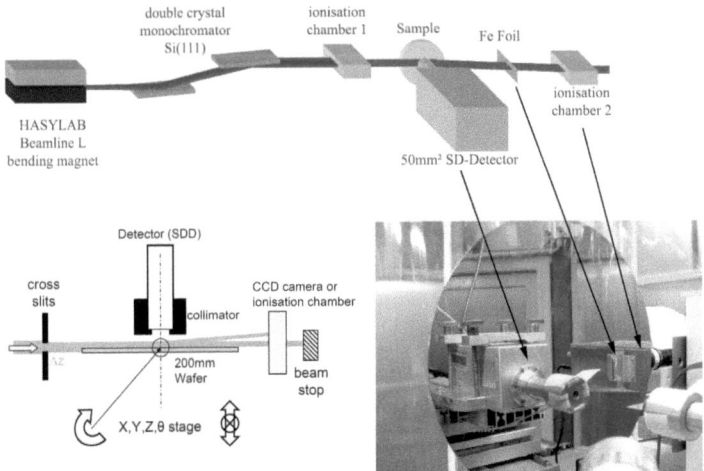

Figure 4.1: Experimental setup at the Beamline L at HASYLAB.

For the XANES measurements the excitation energy was tuned from 6950 eV to 7800 eV in varying steps (10 eV to 0.5 eV) across the iron K-edge at 7111 eV. The acquisition time for each spectrum was set to 5 seconds for sample position P5 and P7 and 8 seconds for sample position P21 (see figure 4.3). Including the time needed for motor movements during the scan the overall time for one scan (454 spectra) accumulated to ~50 minutes for P5 and P7 and 75 minutes for P21. For each specimen at least three repetitive scans were performed to increase the signal to noise ratio. During all XANES measurements the absorption of an iron foil was recorded in transmission mode simultaneously. The first inflection point (i.e. the first maximum of the derivative spectrum) of the Fe metal foil scan was assumed to be 7111 eV (Fe-K edge). The energy scale of each XANES scan (standards and samples) was corrected with respect to the Fe-K edge.

The peak fitting of the fluorescence spectra was done within the software package QXAS [116] and the absorption spectra have been analyzed using ATHENA of the IFEFFIT software package [117-119]. ATHENA was also used for the linear combination fitting procedure.

The experimental setup, testing and all measurements were accomplished within 48 hours.

4.3 Results and discussion

Figure 4.2 shows the contamination maps on the 200mm Si wafer obtained with a laboratory TXRF instrument. The TXRF instrument is a Rigaku TXRF 300 3 crystal system with an 18kW generator. The system can measure K line series for elements from Na, Al, Mg through the transition series to heavy elements such as Mo. The strength of the instrument is the full wafer mapping software which was designed as a fast analysis highlighting very high points or repeatable patterns of contamination. There are several papers on the statistics of the analysis [120, 121].

Figure 4.3 shows the Ca and Fe maps of the marked areas (A5, A7, A21) carried out with SR-TXRF. The lateral resolution was matched to the size of the collimator of the detector (8mm diameter) and the vertical beam dimension (1.6mm). The Ca map is shown as an example of how the maps of the other elements were used to double-check that the correct position of interest was found. Due to the limited time frame of the measurements the area scan A5 was stopped after finding a first maximum of iron contaminations. The marked positions (P5, P7, P21) have been chosen for further investigations.

Chapter 4   Silicon Wafer Surface Analysis of Fe contaminations

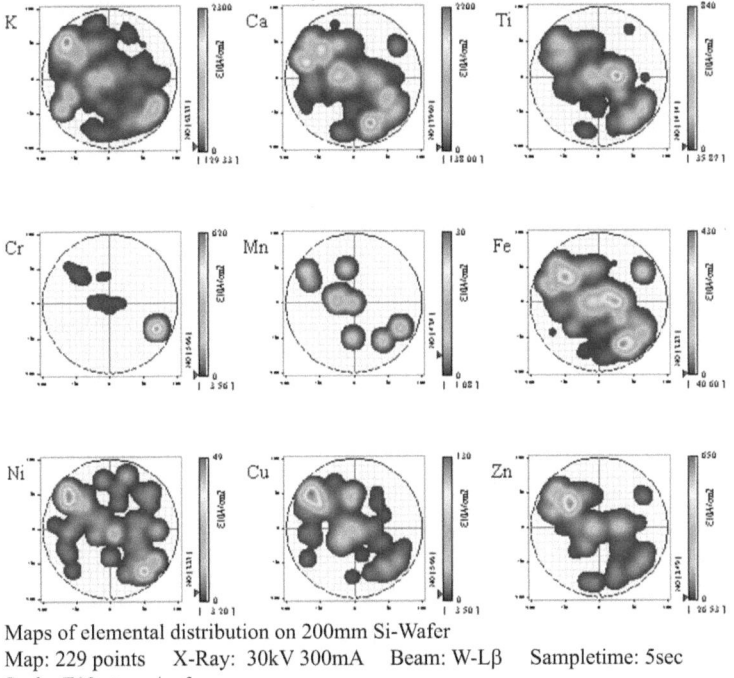

Maps of elemental distribution on 200mm Si-Wafer
Map: 229 points   X-Ray: 30kV 300mA   Beam: W-Lβ   Sampletime: 5sec
Scale: E10 atoms/cm²

Figure 4.2: Maps of the contaminations on the 200mm Wafer measured with a laboratory TXRF wafer-analyzer at IBM laboratories

Angular scans at these positions have shown that all contaminations are present in form of residue on the wafer surface. This is indicated by the typical angular intensity curve [122] of the Fe Kα fluorescence diagramed in figure 4.4.

A spectrum of each point of interest was recorded for 100s. The energy of the incident beam was set to 7800eV for these measurements. The Limit of Detection for a measurement time of 100 seconds (LD100) for Fe was estimated from these measurements to be in the range of 750 fg. The LD was calculated using:

$$LD = \frac{3 \cdot \sqrt{N_B}}{N_N} \cdot m_{sample} \quad (4.1)$$

where N_N and N_B are the net and background intensities of the signal. Here m_{sample} was calculated to ~ 45 pg using the data obtained in the laboratory measurements (~ 400 E10 atoms/cm²) and assuming an inspected area of 1.6 x 8 mm². The LD values for a measurement time of 1000 seconds (LD1000) have been extrapolated to 250 fg. The peak-to-background ratios were found to be 12.6, 9.3 and 3.5 for P5, P21 and P7 respectively.

57

Chapter 4   Silicon Wafer Surface Analysis of Fe contaminations

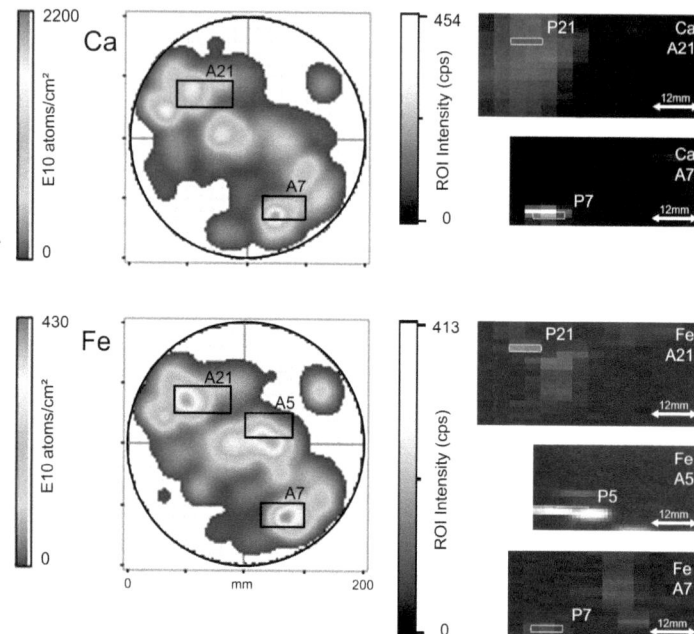

Figure 4.3: Laboratory TXRF maps (left) with marked regions of the areas scanned by means of SR-TXRF (right). The marked regions on the left correspond to the areas on the right. The white boxes on the right localize the points of maximum Fe contamination which were selected for further investigations (P21, P5 and P7).

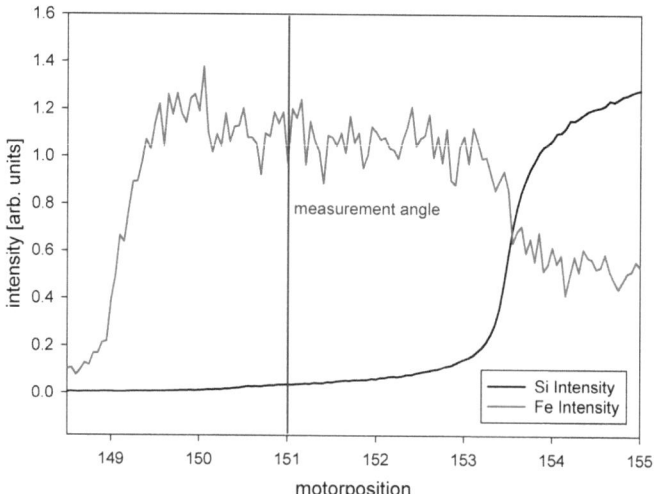

Figure 4.4: Intensity vs. angle of incidence (position of rotator θ) recorded at position P5

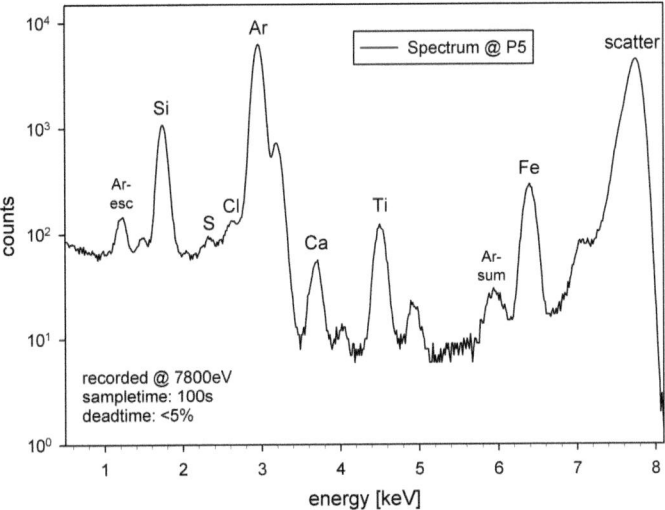

Figure 4.5: Fluorescence spectrum recorded by SR-TXRF for 100 seconds at position P5

Fe K-edge XANES measurements have been carried out in the positions indicated. For each position at least three repetitions were performed. Repetitive scans have been merged together to improve the signal to noise ratio. For the speciation the spectra obtained from the unknown contamination must be compared to spectra of possible reference materials. If the contamination is completely unknown a reliable determination of the compound might not be possible and the investigator must be satisfied with a qualitative indication about the oxidation state of the element examined. Reference samples of common iron species have been measured in vacuum under TXRF conditions [95] for comparison. Figure 4.6 shows all measured Fe standards. The number of standards measured is certainly a key factor in the consideration between overall time of investigation and quality of the results and even success or failure.

Table 4.1 presents the determined edge positions (first inflection point of the XANES spectrum) and the oxidation states of all measured standards and samples. The Fe oxidation states of the samples have been determined by comparison with the standards.

Chapter 4   Silicon Wafer Surface Analysis of Fe contaminations

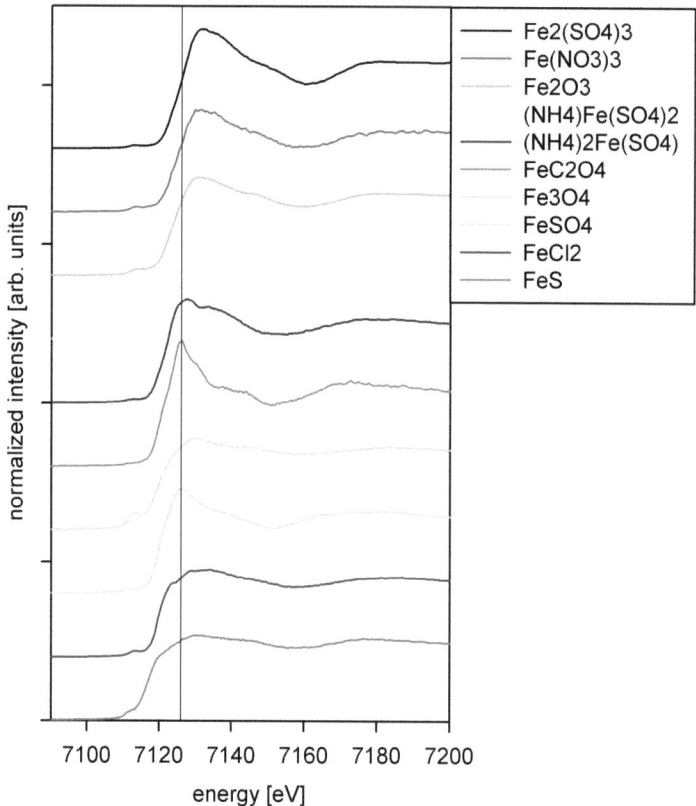

Figure 4.6: All measured Fe-standards. The vertical line at 7126 eV indicates the edge position of the ferric sulfate (Fe2(SO4)3) standard.

	compound	edge position [eV]
FeS	Iron(II)-sulfide	7117
FeCl2	Iron(II)-chloride	7119
FeSO4	Iron(II)-sulfate	7119.5
Fe3O4	Iron(II,III)-oxide	7119.5
FeC2O4	Iron(II)-oxalate	7120.5
(NH4)2Fe(SO4)2	Ammonium-Iron(II)-sulfate	7122.5
NH4Fe(SO4)2	Ammonium-Iron(III)-sulfate	7123
Fe2O3	Iron(III)-oxide	7123.5
Fe(NO3)3	Iron(III)-nitrate	7125
Fe2(SO4)3	Iron(III)-sulfate	7126
Wafer at P5	Fe(III)	7125
Wafer at P7	Fe(II)	7121.5
Wafer at P21	Fe(III)	7124.5

Table 4.1: Oxidations states and edge positions of standards and samples

Figure 4.7 shows the XANES spectra for the three samples along with the spectra of the standards with edge positions closest to the edge positions of the samples.

The XANES spectra were recorded summing all counts within the Fe-Ka Region-Of-Interest (ROI). To make sure that this method is reliable for such low concentrations one XANES scan of each point (P5, P7, P21) was generated using the net-intensity of the Fe-Ka peak. Each fluorescence spectrum of the scan was therefore fitted with QXAS. A comparison of the XANES spectra revealed no significant differences between the scans generated in the two mentioned ways. The apparent discrepancy was the increased signal in the low energy range of the pre-edge region of the scan generated using the ROI method. This effect can be explained by a contribution of the scatter peak increasing the counts of the Fe-Ka ROI at low energies far below the edge. However, as the signal to noise ratio was not improved using the peak fitting method the ROI evaluation was used.

The poor counting statistic for P7 and P21 made the results of further investigations of the Fe-compound on the wafer's surface doubtful. For P5 it was possible to perform linear combination fits to have an estimation of the composition of the Fe contamination. The 3 most appropriate (figure 4.7) standards (Fe2O3, Fe(NO3)3 and Fe2(SO4)3) were used for a linear combination fitting using all possible combinations (four) of these standards. The results and fit parameters of the best fit (best R-factor) are reported in table 4.2. The R-factor was determined using:

$$R = \frac{\sum (data - fit)^2}{\sum (data)^2} \qquad (4.2)$$

	Fe2O3 [%]	Fe(NO3)3 [%]	Fe2(SO4)3 [%]	R-factor	chi-square
P5	26 ±7	48 ±8	26 ±10	0.001424	0.1478

Table 4.2: Results and fit parameters of the best fit for P5.

Figure 4.8 shows the result of the best linear combination fit for sample P5.

Chapter 4   Silicon Wafer Surface Analysis of Fe contaminations

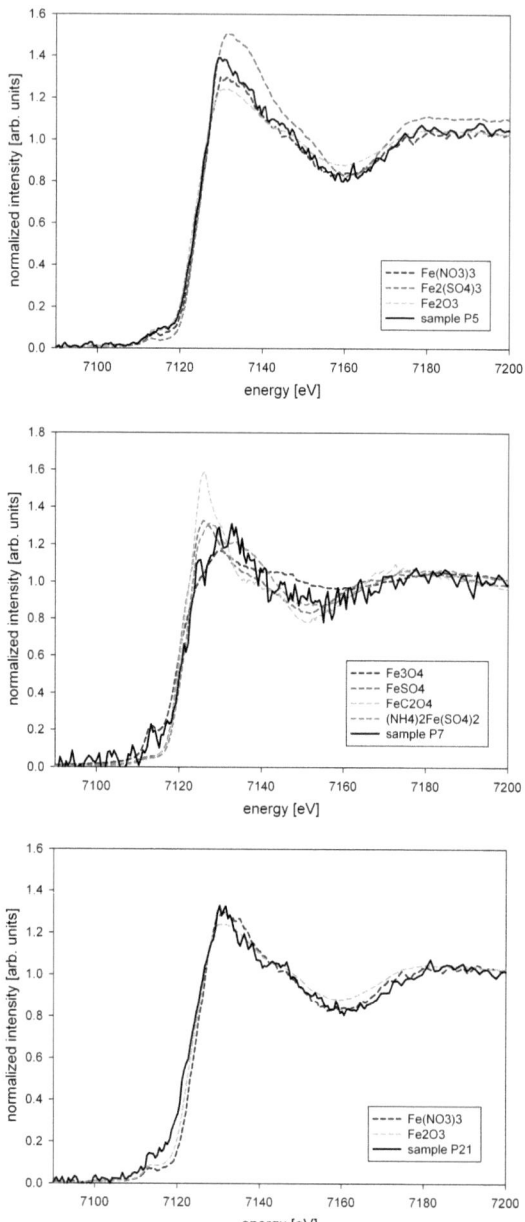

Figure 4.7: XANES scans at positions P5, P7 and P21 in comparison with the spectra of the standards with similar edge positions

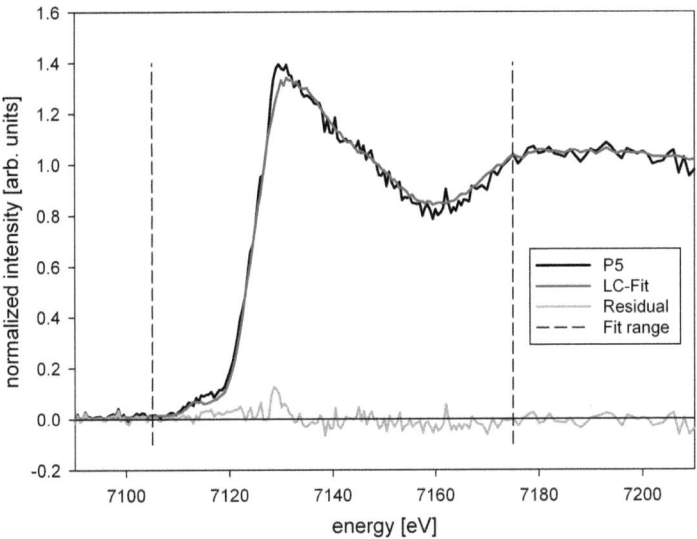

Figure 4.8: Result of the best linear combination fit for sample P5

4.4 Summary

It could be shown that SR-TXRF in combination with XAS enables a XANES analysis of wafer surface contaminations even in the pg region within a reasonable time frame. The setup allows a spatially resolved multi-element analysis of the wafers surface. Additionally the type of the contamination (residual, surface layer, bulk) and the chemical state of a specific element can be determined. All these investigations can be done non-destructively using the same experimental setup. The major problem of this setup is the measurement time and the need for an intense and energy tunable X-ray source. The wafer mapping and XANES scans are time consuming particularly for such low concentrations. The XANES evaluation showed, that a peak to background ratio below 10 for the element of interest in the fluorescence spectra will cause serious problems for the interpretation of the data. Within the time frame of 48 hours 3 points of interest could be investigated. A determination of the oxidation state of iron in the samples was possible. However, for the determination of the iron compound satisfactory results could be achieved for only one point (P5). The evaluation of the other points (P7, P21) with the linear combination fit procedure was not satisfying due to the low Fe concentrations and the limited measurement time resulting in a poor signal to noise ratio of the XANES spectrum. The evaluation of the XAS data revealed that the iron contamination in point P5 is a mixture of iron compounds dominated by iron-nitrate. The presence of iron-nitrate and iron-sulfate on a contaminated wafer shows that several chemical reactions are taking place. The iron-sulfate is particularly surprising and it could be interesting to further investigate how it was formed.

Chapter 5

Characterization of Atmospheric Aerosols

This chapter describes the analysis of atmospheric aerosols performed within the framework of two projects accomplished in cooperation with the following partners:
- Prof. S. Török, Dr. J. Osan and DI. V. Groma from the KFKI Atomic Energy research Institute
- Prof. J. Broekaert and Dr. U. E. A. Fittschen from the Department of Chemistry at the University of Hamburg

The results presented in this chapter have been developed together with the cooperation partners and were published in [123-127].

5.1 Introduction

To understand the effects of aerosols on human health and global climate a detailed understanding of sources, transport, and fate as well as of the physical and chemical properties of atmospheric particles is necessary. An analysis of aerosols should therefore provide information about size and elemental composition of the particles and - if desired - deliver information about the chemical state of a specific element of interest in the particles. The use of a so called impactor allows size resolved sampling of particles according to their aerodynamic diameter. The particles can be directly collected on collector plates which in turn can be used as reflectors for Total Reflection X-Ray Fluorescence (TXRF) analysis (e.g. Silicon wafers, Plexiglass or Quartz carriers). TXRF analysis is especially suitable for the investigation of aerosols because it allows a non-destructive multi-element analysis of trace and ultra trace amounts (ng) of most elements. Multi-element analysis is one of the most adequate tools for deriving fingerprints of different sources of fine particulate matter and hence enables to trace the origin of air pollutants. If Synchrotron Radiation (SR) is used as exciting radiation for TXRF analysis (SR-TXRF) the sampling time (i.e. the time needed to collect the aerosols with the impactor) can be diminished because even smaller sample amounts (in the fg region) can be analysed. This enables a more precise time resolution of atmospheric events. Furthermore, the application of synchrotron radiation allows X-Ray

Absorption Near Edge Structure (XANES) measurements to gain chemical information of a specific element of interest. The oxidation state of elements is of relevance for their environmental impact as the toxicity (e.g. Cr(III)/Cr(VI), As(III)/As(V)) or environmental activity (e.g. Fe(II)/Fe(III)) of an element may differ considerably depending on its oxidation state. Here another advantage of the SR-TXRF analysis is that the aerosols can be measured directly after collecting them on the reflectors surface. An additional sample preparation which could change the chemical environment of the sample is therefore not necessary.

5.1.1 Analysis of airport related aerosol particles with high time resolution using SR-TXRF

This project was done in cooperation with the group of Prof. Török from the KFKI Atomic Energy Research Institute in Budapest, Hungary, and was dealing with the analysis of impactor-collected aerosols to monitor local air pollution.

Urban air quality research is an important part of environmental science, since air quality has a strong effect on human health. In particular the air quality at airport areas is of increasing interest due to the dynamic growth of air traffic. The most critical pollutant here is fine particulate matter ($PM_{2.5}$) - i.e. particles with aerodynamic diameters smaller that 2.5μm - originating from the emission of aircrafts, ground handling vehicles and passenger related cars [128]. There is strong evidence that $PM_{2.5}$ causes more severe health effects than particles with larger aerodynamic diameters [129]. Furthermore this study focused on aerosols with smaller sized particles (0.25-4 μm aerodynamic diameter) because larger particles occur in smaller number concentration (particles/m^3) naturally. To allow statistically significant evaluation of these larger particles longer sampling times would be required. Longer sampling times in turn prevent measurements with a time resolution in the range of aircraft movements. This possibility to perform a multi-element analysis with a high temporal resolution (<20 min) was an important point for the investigation of the aerosols to identify potential particle sources. Therefore short collection times were necessary which required a special instrumentation and a highly sensitive analytical technique. Size fractionated aerosol samples have been collected on silicon wafers using a seven stage May cascade impactor and were analysed by means of SR-TXRF. This combination allowed quantitative determination of ultra-trace concentrations (pg/m^3) of most elements from samples collected for less than 20 minutes, while retaining the full size resolution of the impactor.

The aims of this study were i) to introduce an aerosol analysis method which has potential to be used in industrial/traffic processes where the time scale of the event is similar to the typical sampling duration and ii) to present its suitability for trace element analysis of airport related aerosols.

5.1.2 Characterization of atmospheric aerosols using SR-TXRF and Fe K-edge TXRF-XANES

This project was accomplished in cooperation with the group of Prof. Broekaert from the Department of Chemistry at the University of Hamburg, Germany. Here the elemental composition of atmospheric aerosols was investigated using SR-TXRF and the oxidation state of Fe in the aerosols was determined by means of Fe K-edge XANES in total reflection geometry.

The importance of aerosols not only for human health but also for the cloud formation and the albedo of the earth has become apparent in recent years [130, 131]. The particle size and the elemental composition of aerosols are two important characteristics to determine their toxicity and provide information about their origin and geo potential. Like the project introduced in section 5.1.1 and for the same reasons this study focused on elemental determination according to the particle size with special interest in small aerosol particles - i.e. fine (PM_{10}) and ultra-fine dust ($PM_{2.5}$). Furthermore the research focused on Lead - which is known for its toxicity [132] - and Iron which acts as a micro nutrient for phytoplankton, the basis of marine life. In the European Union a limiting value of 2 $\mu g/m^3$ (24h mean) was defined for Pb. Since the 1970s abatement measures for Pb as additive in fuel, led to reduced levels in the atmospheric environment [133]. Nonetheless Pb is again object of public discussions according to its concentrations in electronic devices and toys. Iron is the most abundant transition metal in atmospheric aerosols and is transported from the great landmasses into the ocean. As plankton is responsible for the Dimethyl sulphide (DMS) production in the oceans and DMS is known to drive cloud formation over the oceans [134] it is also relevant to geo climate. The Iron oxides originating from the earth crust mainly contain Fe(III). For marine organisms the uptake of Fe(II) is much easier than that of Fe(III) and hence a change in the oxidation state is very important. Investigations of redox cycles in the atmosphere indicated that the oxidation state of Fe can be changed due to photo reduction in the atmosphere. Additionally it has been proposed that it can be reduced by sulphur or organic compounds [135]. As Iron undergoes reactions in the atmosphere [136] it is likely that the oxidation state is different for various particle size fractions. To achieve a high particle size resolution a 12-stage round nozzle low pressure Berner impactor covering the size range from 16 to 0.015 μm (aerodynamic particle diameters) was used to collect aerosol particulates in the city of

Hamburg over sampling times up to one hour. The sampling time for the aerosols was kept as short as possible to avoid oxidation of Fe during sampling and to test the feasibility of the method for analysis of aerosols collected during short sample times (<1h). This would enable the detection of changes that may occur due to day and night time and different meteorological conditions in the future.

5.2 Experimental

5.2.1 Sampling in Budapest

Airport related aerosol samples were collected in the inside area of Budapest Ferihegy Airport. The airport is located 15 km south-east from the city centre of Budapest. A highway with approximately 50000 cars/day traffic is located 2 km far from the airport area. First, samples were collected on 20x20 mm^2 silicon wafers using a 7-stage May cascade impactor [137]. The May impactor (figure 5.1 left) has aerodynamic cut-off diameters of 16, 8, 4, 2, 1, 0.5, 0.25 µm for stages 1–7, respectively, at 20 l/min sampling flow rate. The sampling duration took 1 (for stage 7), 5 (for stage 6), 10 (for stage 5) and 20 minutes (for stage 4) to obtain the best loading of particles in the impacted strips. Aerosol collection was performed only for the above specified four stages (4 to 7) because this study focused on the fine aerosol fraction only. The May impactor has an impacting slit at each stage and thus the aerosols collected on a silicon substrate showed a pattern of a thin line with approximate dimensions of 20x0.3 mm. In order to quantify each trace element an internal standard - a 300µm long strip containing 5.75ng Cr - was applied on the Si wafers prior to collection of particulate matter. This strip was placed in the middle centerline of the wafers exactly at the position where particles from the air sample were collected. Two sample sets were collected at the runway and at terminal 2 (high traffic: 8 aircraft movements during sampling time), and one at terminal 1 (low traffic: 1 aircraft movement during sampling time) three days later. This sampling day was a Sunday with much lower traffic at the nearby highway and the city of Budapest.

5.2.2 Sampling in Hamburg

The aerosols were collected in the city of Hamburg on a university building at 10 m height using a 12-stage low pressure Berner impactor (figure 5.1 right) at a flow rate of 1.5 m^3/h. Due to the construction of this type of impactor, the particulate matter is deposited in small spots on the impaction plates. The impactor was cleaned before and after each sampling with distilled water in an ultra sonic bath for one hour. The particles were collected on Si carriers covered with silicone to reduce "bounce off" effects. Two sampling campaigns have been performed. Firstly particles were collected in three size fractions of 10.0-8.0µm, 8.0-2.0µm and 2.0-0.13µm (aerodynamic particle size) during the day with sampling times of 60 (series a) and 20 minutes (series b) and during the night with a sampling time of 60 minutes (series c). These samples were analysed with both SR-TXRF and Fe K-edge TXRF-XANES. In a second experiment particles were collected for 60 minutes during the day but in ten size fractions with aerodynamic particle diameter from 16-8 µm, 8-4 µm, 4-2 µm, 2-1 µm, 1-0.5 µm, 0.5-0.25 µm, 0.25-0.130 µm, 0.130-0.060 µm, 0.060-0.030 µm, 0.030-0.015 µm. These samples were kept under Argon atmosphere and analysed by Fe K-edge TXRF-XANES. For quantification of the samples analysed with SR-TXRF one droplet containing 160 pg of Cobalt (internal standard) was spotted with an HP DeskJet 500C inkjet printer onto the aerosol as described in Ref. [138]. To correct for contaminations blank levels originating from the sampling device and the calibration procedure were studied using a laboratory TXRF instrument (ATOMIKA Extra II). Blank levels of Fe corresponded to 1-10% of Fe in the aerosol samples. Blank levels stemming from the internal standard were found to be negligible. The following Iron compounds were used as a reference for XANES measurements: Iron(II)-sulphide, Iron(II)-chloride, Iron(II)-sulphate, Iron(II,III)-oxide, Iron(II)-oxalate, Ammonium-Iron(II)-sulphate, Ammonium-Iron(III)-sulphate, Iron(III)-oxide, Iron(III)-nitrate, and Iron(III)-sulphate (Merck). These reference compounds were suspended in Isopropanol (Merck) and pipetted on the reflectors.

Figure 5.1: Left: 7-stage May cascade impactor and sample collection plate with 20x20 mm² silicon wafer. Right: Low pressure Berner impactor with a modified impaction plate carrying four Si-wafer plates which were directly used as sample carrier in SR-TXRF analysis.

5.2.3 Measurements

The SR-TXRF measurements were performed using the TXRF set-up comprising a vacuum chamber at HASYLAB bending magnet beamline L [95], recently equipped with a sample changer capable to carry eight samples [139]. This allows measuring 8 samples consecutively without the necessity to close the beam shutter and open the vacuum chamber, resulting in a significant time saving during the beam time. The exciting synchrotron radiation was monochromized to 18.5 keV with the aid of the NiC multilayer monochromator. The samples were mounted vertically in front of the side-looking detector and the beam dimensions were set to 200µm x 1400µm (horizontal x vertical). A peltier-cooled silicon drift detector (SDD) with an active area of 50mm² (VORTEX 50 mm², Radiant Detector Technologies) [114, 115] was used to collect the fluorescence radiation. To improve the performance of the TXRF analysis several special detector collimators were developed (an example is shown in figure 5.2). The used materials and the composition were chosen to guarantee optimized absorption of unwanted scattered radiation of the primary beam and its higher harmonics. The sandwich construction furthermore prevented detection of the fluorescence radiation emitted by the collimator materials themselves excited by the scattered radiation and the fluorescence radiation of the samples. The design was optimized for excitation energies of 17keV. The used material were Plexiglas (carrier), Silver (99.95% purity), Tungsten (99.95% purity) and the compound AlMg3. Pure Aluminium (99.999% purity) – as indicated in figure 5.2 – had to be replaced by AlMg3 because it turned out to be too soft for processing.

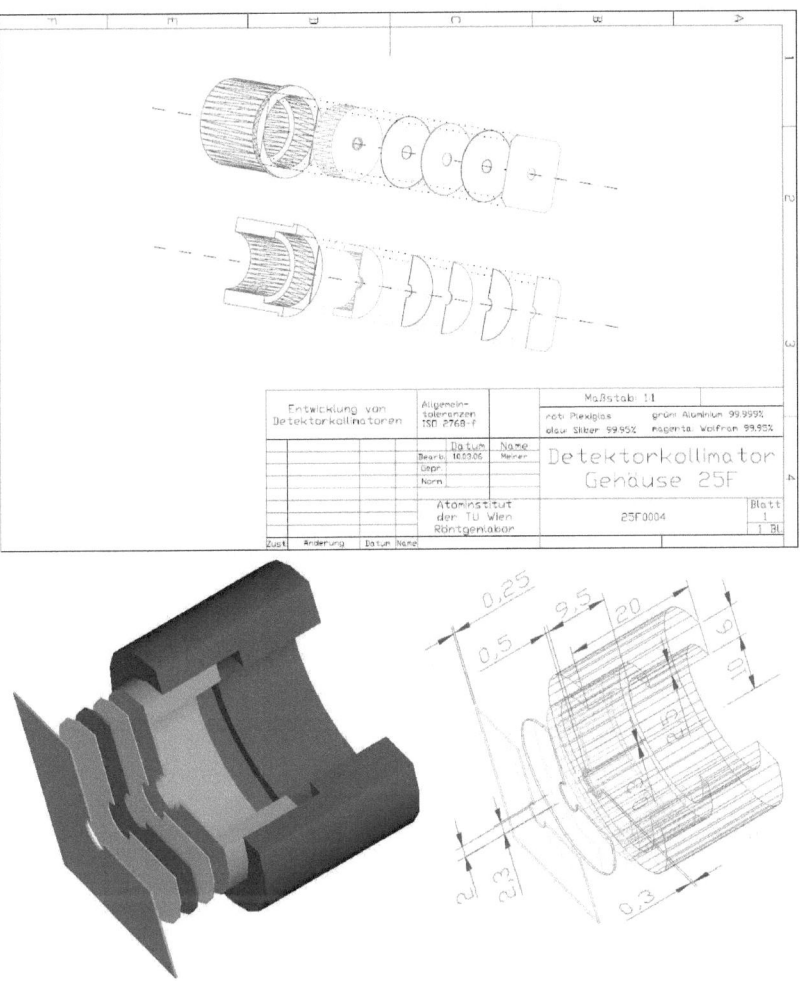

Figure 5.2: Design of the collimators developed for TXRF analysis. Different colors indicate different materials: red: Plexiglas; blue and yellow: pure Ag (99.95%) thickness: 100µm; magenta: pure W (99.95%) thickness: 300µm; green: AlMg3 for the carrier (instead of pure Al; see section 5.2.3) and pure Al (99.999%) for the concluding plate with a thickness of 250µm. The total thickness of the collimator in front of the detector window was 1.05mm minimizing the distance between sample and detector.

A set of pinhole collimators with different pinhole diameters (2, 4 and 8mm) has been produced as well as one slit collimator (1.5 x 8mm; see section 5.2.3.1).

Due to the diverse sample shapes of the collected aerosols (May impactor: strips, Berner impactor: spots) different measurement procedures were performed:

5.2.3.1 Aerosols collected with May impactor (at Budapest Airport)

The SD detector in use was equipped with a special collimator with a 1500μm wide slit made of Mo, in order to fit to the geometry of the impacted aerosol samples. Due to the active area of the SDD (50mm²) all X-Ray photons emerging from the total 20 mm length of the aerosol strip could not be detected. Therefore the homogeneity of the aerosol deposition had to be tested to make sure that a representative part of the strip was measured afterwards. For the test scans, the sample strip and the collimator slit were mounted perpendicular to the X-ray beam, therefore only a 1400 μm section of the sample strip was excited. Here the geometry of the detector collimator matched that of the sample hence this setup resulted in a shielding of the off-lier contaminants as the slit prevented their detection. The vertical scans were performed by moving the sample–detector system relative to the position of the beam using a stepping motor. The step size was set to 1000 μm, and the individual spectra were collected for 3 seconds. The measurements performed to determine the elemental concentrations were carried out applying the SR beam parallel to the sample strip (this means the sample was rotated by 90° in comparison to the first setup). The position of the sample strip was determined using vertical scans with 200 μm step size. Individual spectra were collected at the strip positions for 100 seconds. The net characteristic X-ray intensities of the elements were calculated by evaluating the spectra by non-linear least-squares fitting using the QXAS (Quantitative X-ray Analysis System) software package [116]. The quantification of the elements found in the samples was performed using the fundamental parameter method [140].

5.2.3.2 Aerosols collected with Berner impactor (in the city of Hamburg)

Here a pinhole collimator with 4 mm diameter was used in front of the detector. Vertical line scans with 200 μm step width have been performed in order to localize the collected aerosol and the applied picodroplet containing the Co standard. The measuring time was set to 100 seconds for each step. To assure an accurate quantification the single spectra of each line scan have been summed to make sure that the total amounts of aerosol and standard were measured respectively. This was necessary because the applied Cobalt picodroplet was not always exactly at the position of the aerosol spot. The Ka-lines were used for evaluation of all elements investigated except for Pb for which the La-lines were used. The amounts were calculated from fundamental parameters [140]. The same set up but with a different monochromator was utilized to carry out K-edge XANES measurements for Fe in the fluorescence mode. The exciting energy was tuned with a Si(111) double crystal monochromator from 6950 eV to 7600 eV in varying steps (10 eV far above the edge to

Chapter 5   Characterization of Atmospheric Aerosols

0.5 eV at the absorption edge) across the Iron K-edge at 7111 eV. The fluorescence spectra were recorded using the silicon drift detector described above. The acquisition time for the spectra of a scan was set from 2 to 10 seconds depending on the sample mass to get a reasonable signal to noise ratio. For each specimen at least three repetitive scans were performed to improve the measurement statistic. To check the energy calibration the absorption by an elemental Fe foil was recorded in transmission mode simultaneously for each scan. The first inflection point (i.e. the first maximum of the derivative spectrum) of the Fe metal foil scan was assumed to be 7111 eV (Fe-K edge). The energy scale of each XANES scan (standards and samples) was corrected with respect to the Fe-K edge. The absorption spectra have been analyzed using ATHENA of the IFEFFIT software package [117-119].

5.3 Results and discussion

5.3.1 Results of the analysis of airport related aerosol particles

5.3.1.1 Sample homogeneity

As already mentioned in section 5.2.3.1 is was important to test the homogeneity of the aerosol deposition to make sure that the measured sections of the aerosol strip are representative for the total length and thus the calculation of the total mass deposited here. If this is proven, the extrapolation from the mass calculated for the measured section to the total mass deposition is permitted. The total mass deposited represents the aerosol concentration (ng/m^3) in the collected air volume. For this purpose 18 mm of the existing 20 mm strips were scanned in 1mm steps. As an example the intensities of Fe - which is an indicator element for the aerosol - are shown in figure 5.3. The intensities have been normalized to 100mA ring current and vary within 10% relative standard deviation during the scan, supporting the proposed extrapolation.

Chapter 5   Characterization of Atmospheric Aerosols

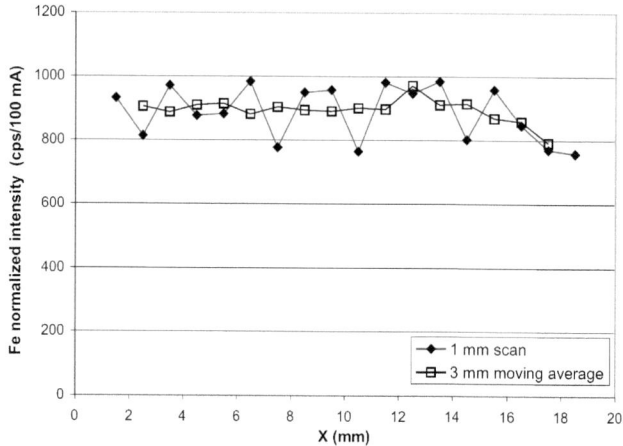

Figure 5.3: Results of the homogeneity test of the 20mm aerosol strip. Iron intensities were chosen because they are representative for the whole aerosol.

5.3.1.2 Multi-element analysis

Elemental concentrations in the samples were calculated based on SR-TXRF spectra (collected for 100 s and normalized to 100mA ring current) using the Cr strip as internal standard. A typical spectrum is presented in figure 5.4. Detection limits achieved for 20 minute sampling time are ranging from ng/m^3 for the light elements (Al, Si) to pg/m^3 for the medium Z elements like Rb and Sr in the present matrix (see table 5.1). The limits of detection (LD) were calculated for 100 s measuring time (LD100).

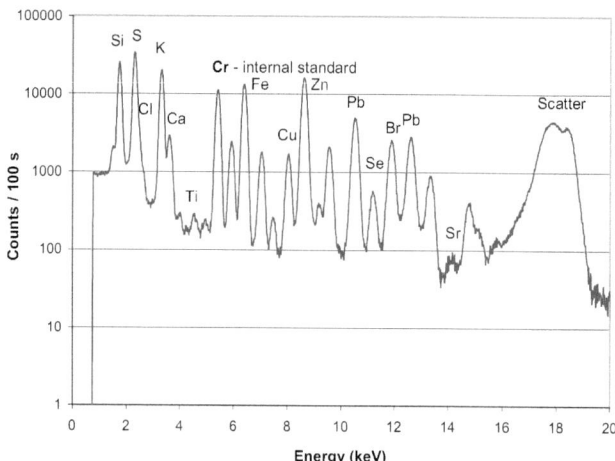

Figure 5.4: Typical SR-TXRF spectrum of an aerosol sample (size fraction 0.5-1 µm) collected at the Budapest airport

74

Chapter 5 Characterization of Atmospheric Aerosols

Element	LD100 (pg/m^3)
S	451.3
Cl	282.8
K	107.9
Ca	70.2
Ti	48.7
Cr	23.4
Fe	12.4
Cu	4.5
Zn	3.5
Se	2.6
Br	2.4
Sr	3.4
Pb	5.3

Table 5.1: Detection limits for 100s measuring time (LD100) calculated for each element detected in the aerosol particles collected at airport sites.

The results of the evaluation of the SR-TXRF measurement of airport-related aerosols sampled with the May impactor are presented in figures 5.5 and 5.6. Since aerosols were collected in 4 different size fractions at 3 different sites, measurement results for 12 samples are presented. Each diagram is relevant for one of the elements studied (S, Cl, K, Ca, Ti, Fe, Cu, Zn, Se, Br, Sr, Pb). Each column represents a measurement site: runway, terminal 2(high traffic) and terminal 1 (low traffic). The size distribution of the concentrations for a specific element is presented within the columns. Diagrams are scaled to the maximum values of the different elements. The results of the measurements shown in figures 5.5 and 5.6 confirm facts which were previously expected for typical suburban areas, besides results which show specialties of airport related aerosols. Aerosols originate from long range transport, city plume and from local sources. Local sources show large variability in the airport, since aircrafts, ground supporting vehicles and also passenger cars have to be taken into account. The emission characteristics are all different, some of them are well known, but the chemical composition of some particles (especially the aircraft related) is still unknown. The results can be discussed in 3 different aspects: (i) magnitude of concentration values at different sites, (ii) typical elements at different sites and (iii) size distribution.

The most obvious fact – which was of course expected - is that the concentrations of all elements measured at the runway and terminal building with larger traffic (terminal 2) are much higher than values at the terminal with lower traffic (terminal 1). Airport traffic here does not only mean movement of aircrafts but also movement of ground supporting and passenger cars which both show correlation with aircraft movements. To be able to determine the origin of the aerosol, samples collected at different sites need to be compared. Since the source properties at terminal buildings are complex (large temporal and spatial variation) it is not possible to determine the origin of particles from their chemical composition. Results from terminal building compared to results from the runway allow differentiating between the sources.

Based on measurements at several European sites it could be shown that Sulfate is dominant in smaller size fractions - its size distribution shows a maximum at ~0.3 µm for rural and ~0.7 µm for urban aerosols [141, 142]. This is in good agreement with the results shown in figure 5.5a. The highest Sulfur concentration was found in the 0.5–1 µm fraction at the runway and at terminal 2, and in the 0.25–0.5 µm fraction at terminal 1. This means that the size distribution of Sulfur has a more rural characteristic for low traffic conditions, and more urban at sites influenced by high traffic.

In figure 5.5c it can be seen that the size distribution of Potassium is very similar to that of sulfur. High concentration values of potassium usually occur near wood-burning emission sources which cannot be found inside the airport area but probably somewhere outside, near the airport. This is confirmed by the fact that Sulfur was also connected to wood-burning processes in previous studies [143].

Significant amount of Chlorine was found to be present in the samples. Besides long-range transport of sea salt, chlorine also originates from the salt used for preventing ice formation on roads during winter time (the sampling was done in January). The results are in accordance with this since a decrease in traffic at the airport has less influence on the concentration of Chlorine than on the concentration of other elements (see figure 5.5b).

Figure 5.5d-f shows that Ca, Ti and Fe have a similar size distribution at the three locations. Furthermore the total concentration values of these elements have similar ratio among the three sites. This leads to the conclusion that the source of particles containing these elements is the same. These elements dominate in larger particulates and therefore they could originate from resuspension (by movements of vehicles and the blowing effect of the aircraft engine).

Chapter 5   Characterization of Atmospheric Aerosols

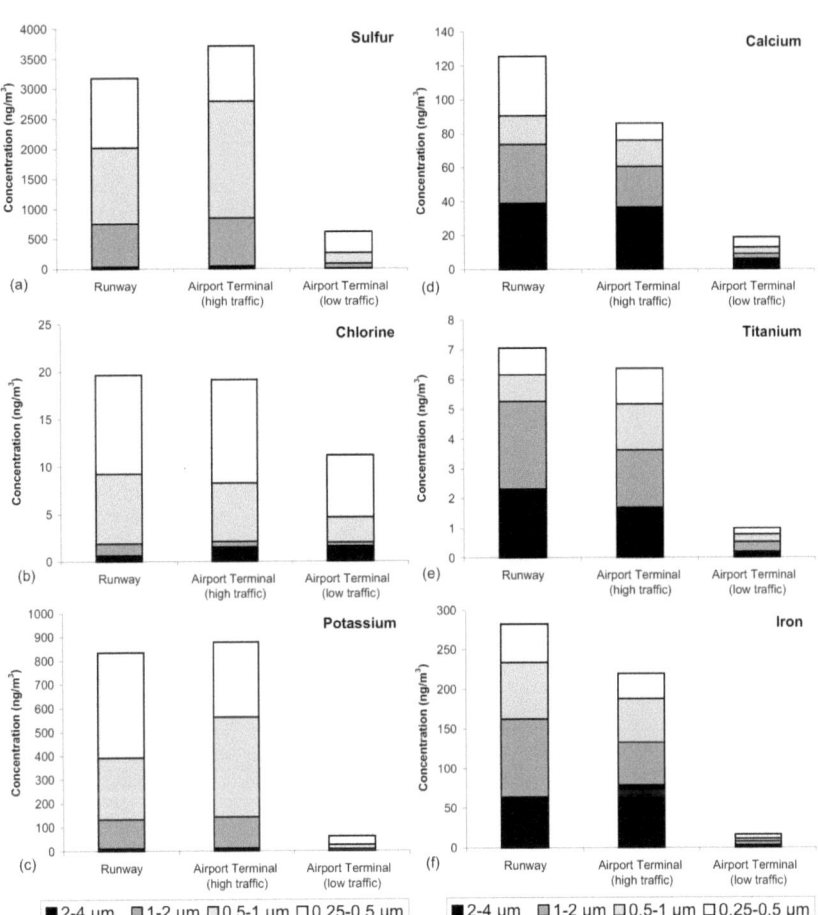

Figure 5.5: Size distribution of elemental concentrations for S, Cl, K, Ca, Ti, and Fe at different sampling sites

Chapter 5   Characterization of Atmospheric Aerosols

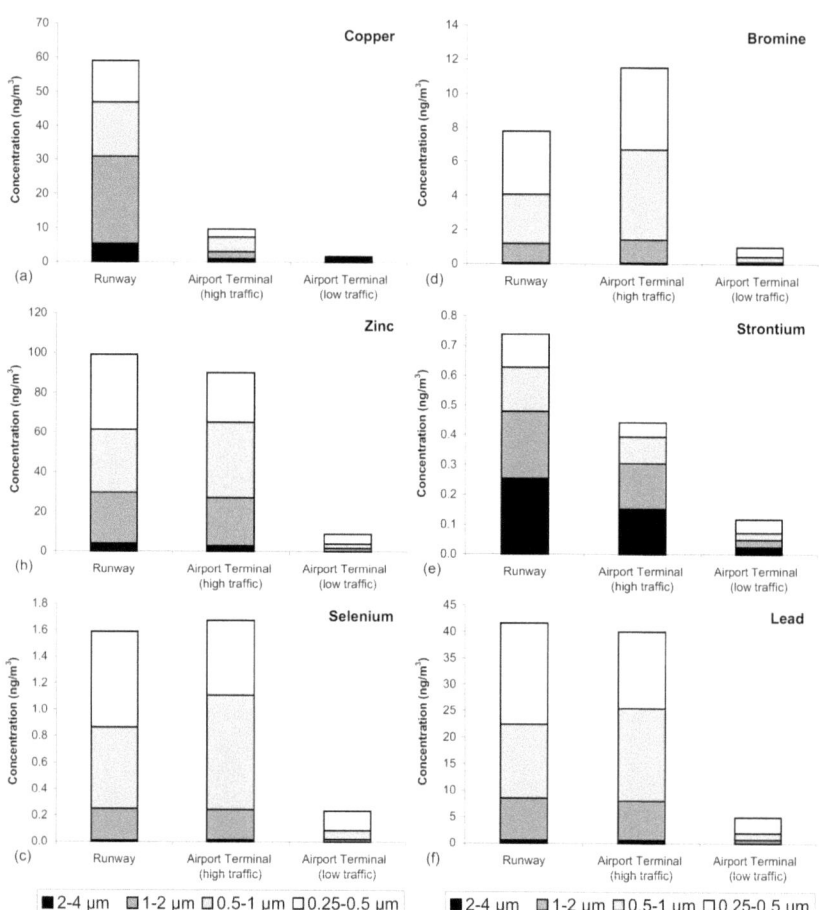

Figure 5.6: Size distribution of elemental concentrations for C, Zn, Se, Br, Sr, and Lead at different sampling sites

Zinc and Lead (see figure 5.6b and f) are usually related to engine combustion processes [144]. Traffic related particles containing these elements usually occur in fine aerosols, which is in good agreement with the results. The concentration values of these elements show similar magnitudes: high at high traffic periods and low at the time of low traffic, but no significant difference could be seen at the runway and terminal 2. Therefore it can be concluded that particulates containing these elements originate from the airport related ground supporting vehicles as well as from the city traffic, transported to airport area.

The only element that shows significant concentration differences between terminal and runway areas is Copper (see figure 5.6a). Copper rich particles probably originate from aircraft brake pad erosion. No other studies have been done connected to the erosion during aircraft landing, which means further investigations are needed.

Strontium is a chemical substitute for Calcium and therefore the size distribution and the Sr/Ca ratio at different sampling sites should be the same. This is in good agreement with the results (see figures 5.5d and 5.6e) although Sr showed the lowest concentration of all elements (sub-ng/m^3). This very good correlation between Sr and Ca concentrations in all size ranges clearly shows the suitability of SR-TXRF for trace element analysis of aerosol samples.

Selenium and Bromine rich particles were found to be dominant in the sub-micrometer size fractions (figure 5.6c and d). Therefore these trace elements can be connected to the combusted fuels used either in air or in ground traffic. The detection of such small quantities in the small size fractions confirms the fact that this method opens up new opportunities in aerosol analysis research.

5.3.2 Results of the characterization of atmospheric aerosols collected in the city of Hamburg

During the first campaign atmospheric aerosols were collected for 20 and 60 minutes during daytime and 60 minutes during the night in the city centre of Hamburg. Fe K-edge TXRF-XANES on Fe were performed prior to the TXRF analysis, because the application of the reference element (internal standard) necessary for quantification in SR-TXRF could have changed the Fe species.

5.3.2.1 Multi-element analysis

The quantification of the elements present in the samples was based on the SR-TXRF spectra (100 s life time, corrected for dead time and normalized to 100mA ring current) using the applied Co droplet as internal standard. A typical spectrum is presented in figure 5.7. Detection limits for 100 s measuring time (LD100) were calculated for all size fractions for Lead an Iron. The LD100 mean value of all size fractions for Fe was found to be 11.5 pg/m^3 (hourly sampling) which is in agreement with the results presented in section 5.3.1.2. However, to estimate detectable levels of Fe in the aerosol particulates, blank values stemming from the impactor have to be taken into account. Those were found to be in the range of 5-17 pg. When considering these blank values the mean detectable level of Fe in the aerosols is much worse than calculated before and corresponds to 245 pg/m^3. Detection limits for Pb (where no blank values had to be considered) were found to be in the range from 1 to 17 pg/m^3 depending on the size fraction. From these values the LD100 mean value of all size fractions for Pb was calculated to 10 pg/m^3 (hourly sampling).

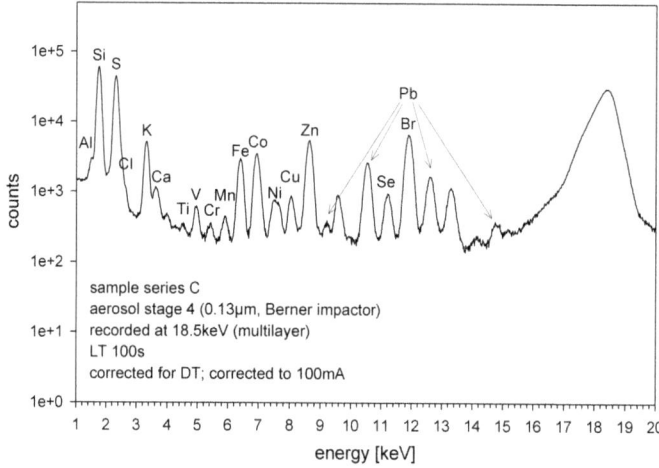

Figure 5.7: Typical SR-TXRF spectrum of an aerosol sample (size fraction 0.13-2 µm) collected in the city centre of Hamburg

The results from the SR-TXRF analysis of the collected aerosols are presented in figure 5.8. It can be seen that 20 min of sampling time gave still enough sample material for elemental determination of most elements. The results for the different size fractions show significant differences between the elemental compositions, as could be expected according to the studies of reference [145]. Furthermore the results from the first (daytime) and the last 60 min sampling (during the night) vary in the composition which indicates a variation of aerosol composition in the atmosphere between the two sample periods. As the quantification is based on single measurements no statistical uncertainties have been calculated.

It was already discussed in section 5.3.1.2 that traffic related particles containing Zinc and Lead usually occur in fine aerosols, which is in good agreement with the results shown in figure 5.9. Series c (sampled during night time) shows lower concentrations of these elements approving the fact that Zinc and Lead can be related to city traffic. Summation of Pb concentrations over all stages gives bulk elemental concentrations of 5 ng/m^3 for series "a" and 2 ng/m^3 for series "c". These concentrations are very similar to those presented in section 5.3.1.2 (low traffic period) and in the same order of magnitude as the concentrations (1.6-90 ng/m^3) determined during a large-scale campaign on post-abatement Pb-levels in several

Chapter 5   Characterization of Atmospheric Aerosols

urban and rural sites in Germany. In this campaign the aerosols were collected over a time period of 24h [146].

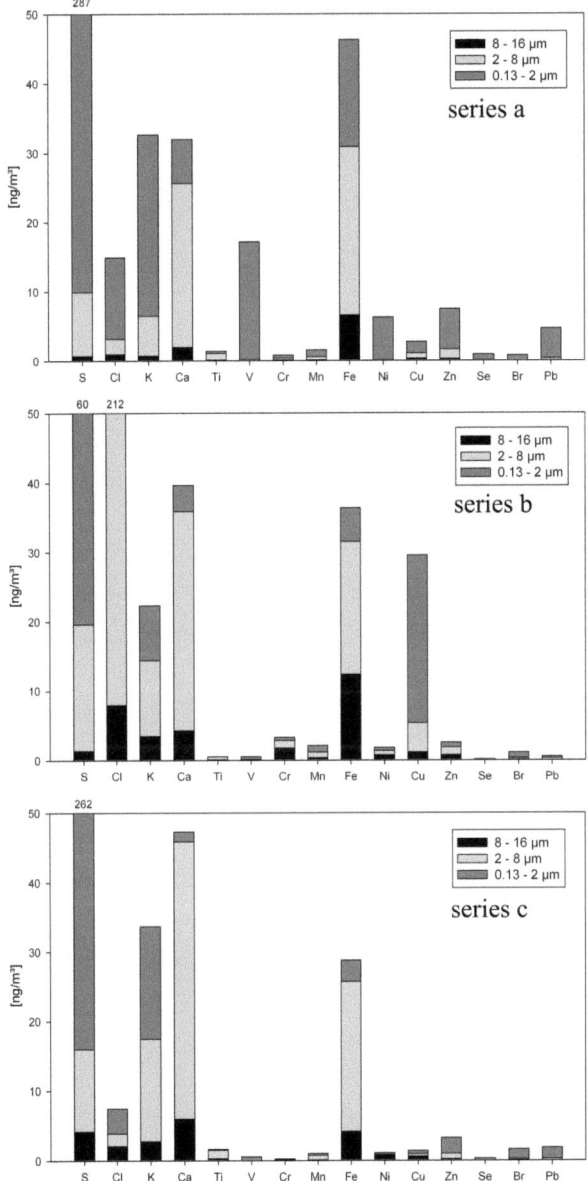

Figure 5.8: Size distribution of all elemental concentrations determined in atmospheric aerosols from three different samplings: 60 min day time (series a) 20 min day time (series b) and 60 min night time (series c)

Chapter 5   Characterization of Atmospheric Aerosols

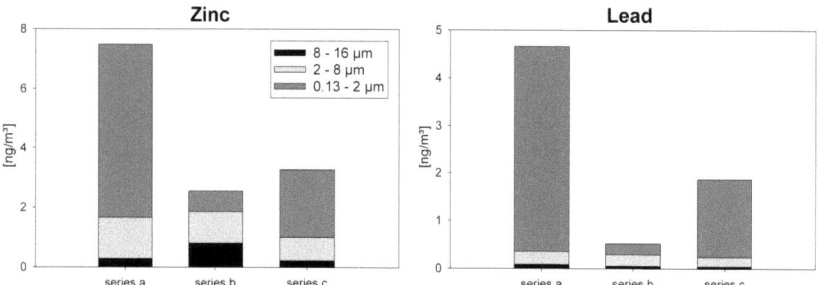

Figure 5.9: Size distribution of elemental concentrations of Zn and Pb evaluated for aerosols collected for 60 and 20 min during the day (series a and b respectively) and 60 min during night time (series c)

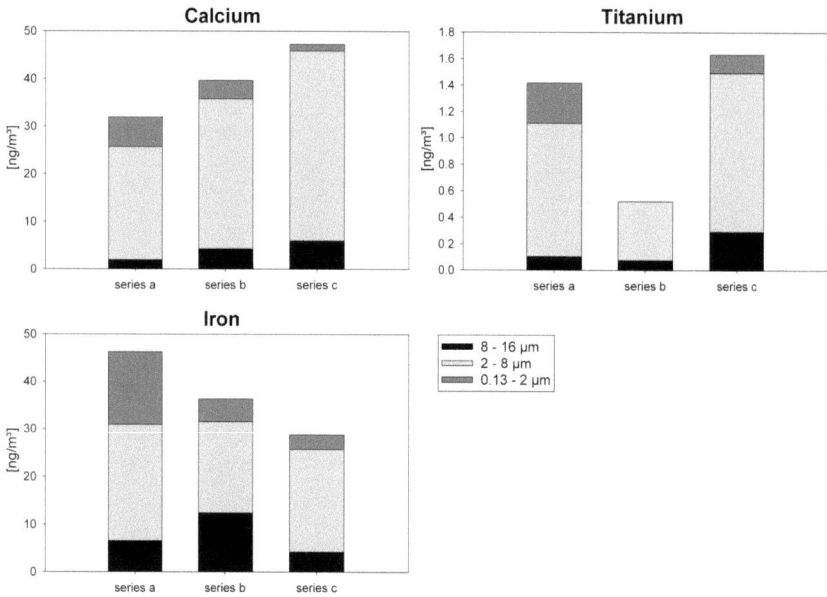

Figure 5.10: As figure 5.9 but for elemental concentrations of Calcium, Titanium and Iron

The results for Ca, Ti and Fe shown in figure 5.10 are also in accordance with the considerations of section 5.3.1.2. These elements have a similar size distribution for all three series and dominate in larger particulates – therefore it is likely that they originate from resuspension processes.

Chapter 5 Characterization of Atmospheric Aerosols

5.3.2.2 Iron K-edge TXRF-XANES analysis

To learn more about the redox characteristics of Iron present in atmospheric particulate matter its oxidation state was analysed by Fe K-edge TXRF-XANES with respect to the particle size. As several studies reported, Iron is photo chemically reduced in the atmosphere hence it seemed possible that Fe(II) was enriched in specific particle size fractions according to its generation.

Different Fe salts analyzed as reference samples are shown in figure 5.11. The references were prepared on silicon reflectors from low concentrated suspension of different Fe salts in isopropanol to avoid oxidation which more easily occurs in aqueous solutions. All spectra have been energy calibrated and normalized as described in section 5.2.3.2. The measured XANES match very well with those found in literature [44] showing the suitability of the procedure used to prepare reference samples from the different Fe species for TXRF-XANES analysis.

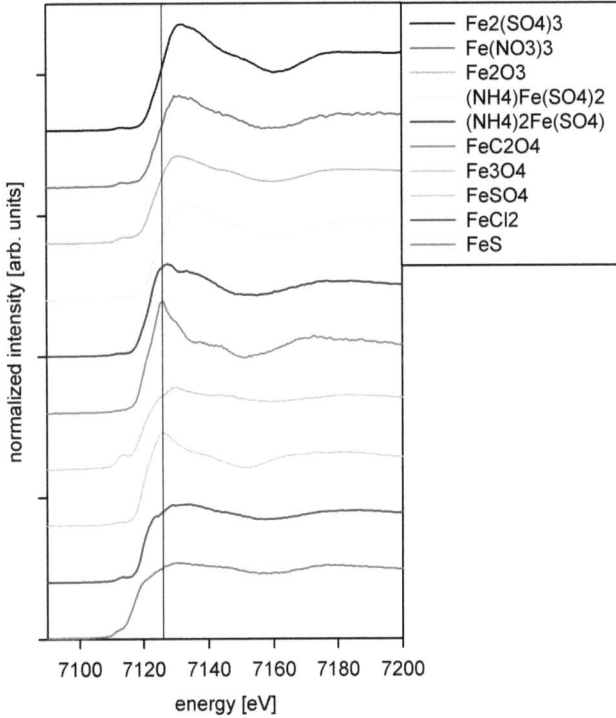

Figure 5.11: Fe K-edge XANES of all measured standards recorded in TXRF geometry. The spectra are displaced vertically for clarity. The vertical line at 7126 eV indicates the edge position of the ferric sulfate (Fe2(SO4)3) standard.

83

According to the XANES analysis of the aerosol collected during the first campaign Fe was present in the oxidation state (III) in all four particle size fractions. Nonetheless it could not be excluded that small amounts of Fe(II) were present but could not be detected because of the dominance of the Fe(III) species. Therefore in a second campaign aerosol particles were collected from ten size fractions. Theses samples were kept under Ar atmosphere to prevent oxidation. Shortly before measurement in the vacuum chamber they were taken out of the protective atmosphere.

Total amounts of Fe found on the collection plates were low (500-900 pg) but significantly higher than the blank values (5-17 pg). These low concentrations and the reduced intensity of the exciting radiation due to the use of the Si(111) monochromator lead to a lower fluorescence intensity. To improve the signal to noise ratio of the XANES spectra repetitive energy scans of the same sample have been merged. Furthermore these scans were used to double check if any oxidation of Fe occurred during the measurements (which were all performed in vacuum). The repetitive scans showed no changes of the fine structure confirming that no oxidation of the samples took place during the measurements.

Figure 5.12 shows the results of the XANES analysis. It is obvious that the XANES spectra of all size fractions were identical (within counting statistics). Furthermore the absorption edges of all aerosol spectra were found exactly at the same energy position as the absorption edge of the Iron(III)-oxide (Fe2O3) standard. Therefore it was concluded that all fractions contain Fe predominantly in the oxidation state of three. This is in good agreement with other studies which all found predominately Fe(III) in the aerosol particulates [147-149].

Figure 5.13 shows a comparison of the XANES spectra recorded for the 8-16 µm size fraction (as a representative example for all fractions) and the Fe2O3 reference. It can bee seen that the pre-edge peak, the white line (highest maximum), the shoulder after the white line, the first minimum and the following maxima and minima of the absorption fine structure are at the same energy positions for both spectra. Differences concern the heights of the white line and the first minimum. This damping of the oscillations of the higher concentrated standard sample could be explained by a self-absorption effect due to the TXRF geometry. Here the path length of the incident beam in the sample is longer than in other geometries and therefore its absorption in the sample cannot be ignored for larger amounts of concentrated samples. This effect will be discussed in more detail in the following chapters. Due to these differences an exact determination of the Iron compound in the aerosols was not possible but it is very likely that Fe2O3 is the predominant form. The results of this study will allow choosing a specific set of reference samples to perform a reliable speciation of the Iron compound (or the mixture of compounds) found in the particles.

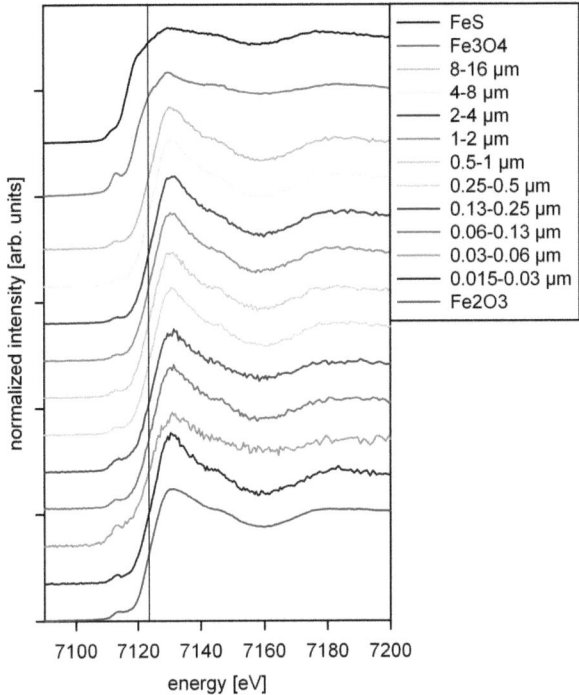

Figure 5.12: Fe K-Edge TXRF-XANES spectra from ten aerosol particle size fractions. Additionally the XANES of the reference standards Iron(II)-sulfide (FeS), Iron(II,III)-oxide (Fe3O4) and Iron(III)-oxide (Fe2O3) are shown for comparison. The vertical line at 7123.5 eV indicates the edge position of the Iron(III)-oxide standard.

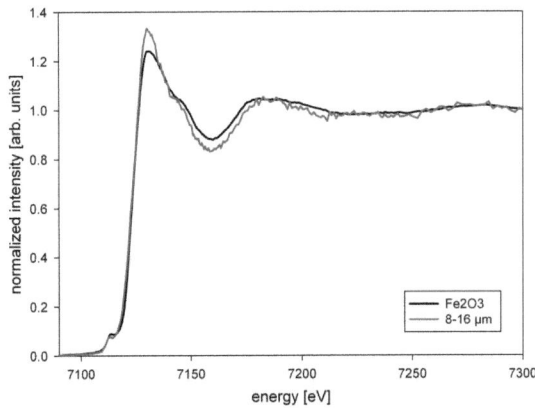

Figure 5.13: XANES spectra recorded for the 8-16 µm aerosol particle size fraction and the Fe2O3 reference

5.4 Summary

Using SR-TXRF elemental amounts in atmospheric aerosols were determined with respect to the particle size. Due to the excellent features of SR-TXRF detection limits in the range of pg/m^3 could be achieved for 20 minute sampling time. This short time collection allows studying temporal variation of elemental concentrations in size-fractioned aerosols. The results have proven the fact that the most important elements occur in concentrations above detection limits. To address the sources of post-abatement levels of Pb in West Europe, information about the variation of Pb in the atmosphere on short time scales and information on Pb concentration according to the particle sizes are necessary. The detection limits found here will allow for a large-scale study of Pb in the atmosphere and offer the possibility of further speciation e.g. by Pb L-edge XANES. The total amounts of Pb (summed over all size fractions) found in both studies are very similar and in agreement with those found in other studies [146]. Evaluation of the size distribution of the elemental concentrations allowed tracing possible particle sources. Lead and Zinc concentrations - which are elements related to combustion processes - showed similar size distributions as well as Ca, Ti and Fe concentrations which could therefore be related to resuspension processes.

5.4.1 Analysis of airport related aerosol particles with high time resolution using SR-TXRF

Aerosol collection with May impactor combined with the SR-TXRF analytical method could be efficiently used to study atmospheric processes on the time scale of minutes. The sampling and trace element analysis was performed at airport sites near sources which show high spatial and temporal variability. Typical aircraft related particles could be identified from extremely small sampling volumes. Special, aircraft related particles were detected near the runway and showed high copper concentrations due to aircraft brake erosion. Size fractioned sampling allowed determining particles not related to airport combustion processes - for example aerosols of larger size fractions containing Ca, Ti and Fe due to resuspension. Furthermore elements originating from wood burning processes like S and K - which are connected to atmospheric transport - could be identified by comparison of concentration magnitudes at different sites.

5.4.2 Fe K-edge TXRF-XANES of atmospheric aerosols collected in the city of Hamburg

The results from the Fe K-edge SR-TXRF-XANES analysis of the aerosol samples showed that mainly Fe(III) was present in all particle size fractions.

The results of the Fe K-edge SR-TXRF-XANES analysis of the aerosol samples showed that Fe was present in the oxidation state of three (predominately in the form of Iron(III)-oxide) in all collected aerosols. This corresponds to other studies on the oxidation state of Fe in aerosols although in rain and cloud samples high amounts of Fe(II) were reported [150, 151]. Possible reasons why exclusively Fe(III) was found in the aerosols analyzed here may be the place or the height where the aerosols have been collected, or some oxidation which may have occurred during the condensation process. Finally it can not be excluded that oxidation took place during storage or sampling, even though the sampling time was kept short and the samples were stored under Argon atmosphere. Further studies will have to focus on these problems.

Chapter 6

Analysis of Arsenic in Cucumber Xylem Sap

6.1 Introduction

Synchrotron Radiation induced Total reflection X-Ray Fluorescence (SR-TXRF) analysis was utilized for X-Ray Absorption Near Edge Structure (XANES) measurements for the speciation of arsenic in cucumber (Cucumis sativus L.) xylem sap. Main objective of this work was to exploit the advantages of the TXRF geometry for XANES analysis of such kind of samples.

6.1.1 Arsenic

The element of interest for these investigations was arsenic because it is a known contaminant of groundwater in the south-eastern part of Hungary. It is of geological origin and can reach concentrations up to 150 ng/mL [152-154]. The World Health Organisation (WHO) recommends an upper limit of 10 ng/mL for arsenic in drinking water [155].

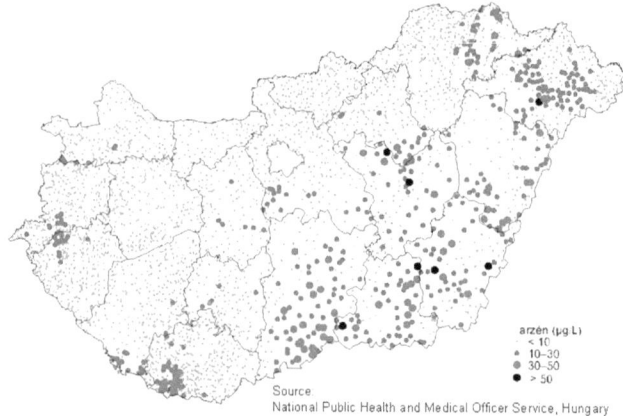

Figure 6.1: Arsenic contamination in Hungarian groundwater.

Chapter 6   Analysis of Arsenic in Cucumber Xylem Sap

Arsenic is a metalloid widely distributed in the earth's crust and occurs in trace quantities in all rock, soil, water and air. It can exist in four valency or oxidation states: –3, 0, +3 and +5. Arsenate (As(V)) is generally the stable form in oxygenated environments like aerobic soils whereas Arsenite (As(III)) is the dominant form under reducing conditions.

The speciation of arsenic is relevant because the toxicity of arsenic differs considerably dependent on the oxidation state and chemical form. Inorganic species, like arsenite (As(III)) and arsenate (As(V)), are more toxic than the organic ones, e.g. monomethyl arsonic (MMA) and dimethyl arsinic (DMA) acids [156, 157]. Arsenite is generally more toxic than arsenate [154, 156, 157] and reacts with sulfhydryl groups of enzymes and tissue proteins, leading to inhibition of cellular function and death [158, 159]. Concerning plants arsenate acts as an analogue of phosphate, competing for the same uptake carriers in the root [153, 158, 160]. It has been shown [153, 154, 157, 161] that plants have the capability to change the oxidation state of arsenic and therefore this issue should be investigated. Research focused on xylem sap, because the plant's xylem is primarily responsible for transportation of water and solutes. Furthermore, only minimal sample preparation was required as the xylem sap can be easily collected and only ancillary filtration has to be done [154, 162].

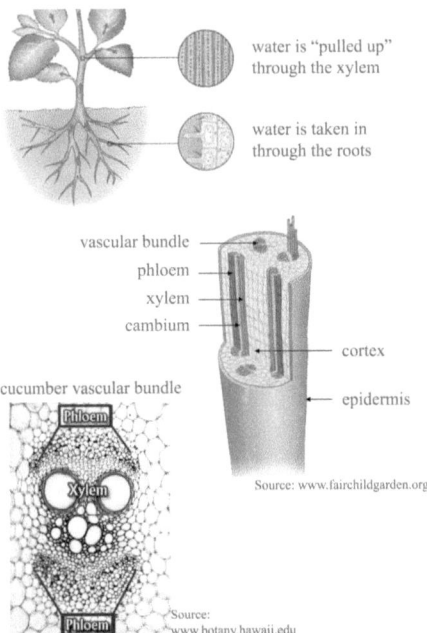

Figure 6.2: Xylem and its role as transport tissue in vascular plants.

89

To understand how (edible) plants metabolise and transform arsenic is essential for mainly two reasons: first of all plants can be used as indicators for the bioavailable part of arsenic in soil and, secondly, the remaining arsenic in plants is available to the next level in the food chain.

The inorganic species arsenate and arsenite are the predominant form of arsenic in terrestrial plants whereas organic species like DMA and MMA have only been found in relatively low concentrations [158]. Therefore nutrient solutions containing arsenite and arsenate and standard samples of these two arsenic species were used.

6.1.2 SR-TXRF

Various publications describing the method of X-ray absorption analysis for the speciation of arsenic in samples with As concentrations in the ppm range can be found in literature [163-167]. However, previous investigation of the cucumber xylem saps with flow injection analysis (FIA) and high performance liquid chromatography high resolution inductively coupled plasma mass spectrometry (HPLC-HR-ICP-MS) revealed arsenic concentrations in the 30-50 ng/mL (ppb) range [154].

As discussed in section 3.3 SR-TXRF offers detection limits in the fg range for transition metals with a multilayer monochromator and a bending magnet beamline [71, 93-95]. If a crystal monochromator is used instead of a multilayer the technique can be extended to X-ray absorption measurements to gain chemical information on a specific element of interest [41, 95, 99-101]. Due to the fact that the flux delivered by a crystal monochromator (e.g. Si(111)) is about two orders of magnitude lower than the one from a multilayer one has a lower sensitivity for X-ray fluorescence analysis. However, this modified setup still offers sufficient sensitivity for elemental analysis at ng/mL (ppb) level. Furthermore it allows the extension of XAS to the trace element level in droplet samples where only small amounts are available [95, 99] and even in the low energy range but using a plain grating monochromator [104]. Therefore the applicability of XANES with TXRF acquisition for the determination of the arsenic species in cucumber xylem saps was tested in this work.

An important point in elemental speciation is to avoid chemical transformation of the samples during the analysis; therefore it is a big advantage if only minimal sample preparation is necessary. SR-TXRF allows analyses of xylem saps directly after collection with micropipettes in argon atmosphere without any further sample preparation. Additionally only few µl of solutions are required for TXRF measurements [71, 93-95, 99] which is another advantage for the analysis of xylem saps.

6.1.3 Aims

The primary motivation for the speciation of arsenic in the xylem of cucumber plants was:
- to understand how plants metabolise and transform As
- to assess the health risk caused by As entering the food chain (different As species have different toxicity)

Previous investigation revealed arsenic concentrations in the 30–50 ng/ml (ppb) range. As these concentrations are too low for a standard X-ray absorption spectroscopy (XAS) setup the
- competitive capability of SR-TXRF-XANES analysis vs. HPLC-HRICP-MS

will be investigated for this application.

6.2 Experimental

6.2.1 Sample preparation

Plant growth and sampling was done at the Plant Physiology Department of Eötvös University of Budapest [154]. Cucumber plant seedlings were grown in a modified Hoagland solution. At two leaf stage the plants have been transferred to nutrient solutions containing 150 ng/mL As(V) or As(III), respectively. After 14 days of this arsenic treatment xylem sap was collected from the stem of the plants, deposited on Quartz reflectors and dried. Prior to the xylem sap collection, plants were kept in arsenic–free nutrient solutions containing double concentration of KNO3 with respect to the primary Hoagland solution for 1 hour in order to enhance bleeding. The stem was cut 2 mm above the root neck and xylem sap was collected with micropipettes from groups of 4 plants for 15 minutes in argon atmosphere and transferred into PE vials immersed in an ice salt bath. The mass of xylem sap collected from plants treated with As(III) and As(V) was determined to be 382 mg and 430 mg respectively.

Standard solutions containing arsenic in concentrations of 10 µg/mL were prepared for both arsenic species. From these solutions standard samples were prepared with different arsenic mass by pipetting 1 µl and 20 µl (4 times 5µl) onto Quartz reflectors. In the case of nutrient solutions and xylem saps, volumes of 10 µl and 20 µl (4 times 5µl) have been applied. After the deposition samples were vacuum dried for 3 to 5 minutes and transported in inert atmosphere (Ar) in order to prevent oxidation.

Mihucz et al. [154] observed a partial oxidation of the arsenite to arsenate in the case of arsenite-containing nutrient solutions. To cross-check this observation samples have been

Chapter 6   Analysis of Arsenic in Cucumber Xylem Sap

taken from the As(III) containing nutrient solutions 48 hours after the plants were placed in these solutions.

Figure 6.3: Left part: Cucumber plant seedlings grown in the modified Hoagland solution. Right part: Vacuum drying of solutions deposited on Quartz reflectors.

6.2.2 Measurements

Arsenic K-Edge XANES measurements in fluorescence mode and grazing incidence geometry were carried out using the setup at the Beamline L at the Hamburger Synchrotronstrahlungslabor (HASYLAB) at DESY [92, 95]. Shortly before measurement the specimen were taken out of the protective atmosphere and placed into the vacuum chamber of the spectrometer. All measurements were performed in vacuum.

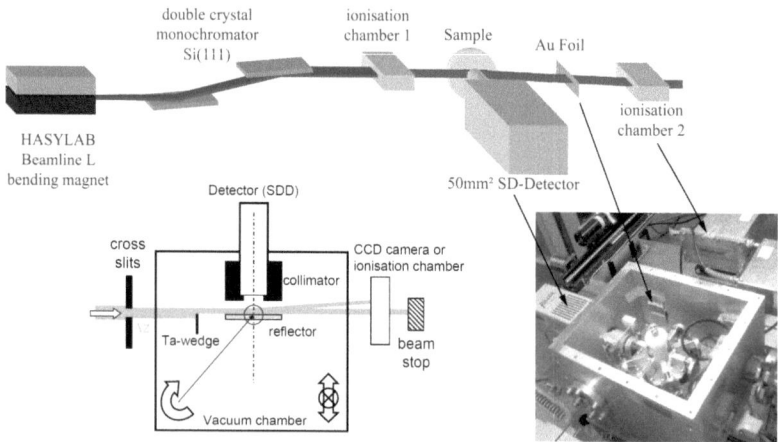

Figure 6.4: Experimental setup at the Beamline L at HASYLAB

A Si(111) double crystal monochromator was used for selecting the energy of the exciting beam from the continuous X-Ray spectrum emitted by the 1.2 Tesla bending magnet at Beamline L. The primary beam was collimated to 200μm x 1400 μm (horizontal x vertical) by a cross-slit system. The incident X-ray intensity was monitored with the aid of an ionization chamber.

During the measurements the excitation energy was tuned in varying steps (5eV to 0.5eV) across the arsenic K-edge at 11862 eV. At each energy a fluorescence spectrum was recorded by a Silicon Drift Detector (SDD, VORTEX 50 mm², Radiant Detector Technologies) [114, 115]. The distance between SDD and sample carrier was 1mm [95]. The acquisition time for each spectrum was set to 5 (or 3) seconds for standard and nutrient solutions and 20 seconds for the xylem sap samples. For each scan 280 spectra have been recorded. For each specimen not less than three repetitive scans were performed.

The critical angle for total reflection changes during an energy scan. In the particular case the critical angle of silicon shifts from 2,67mrad to 2,56mrad for an energy variation from 11700 to 12200eV. On that account the incident angle of the primary X-ray beam was adjusted to 2mrad which is far below the critical angle and it can be assumed that the change of the critical angle during the XANES scans is unproblematic for the measurement of droplet samples (residues on surface) (see figure 6.5).

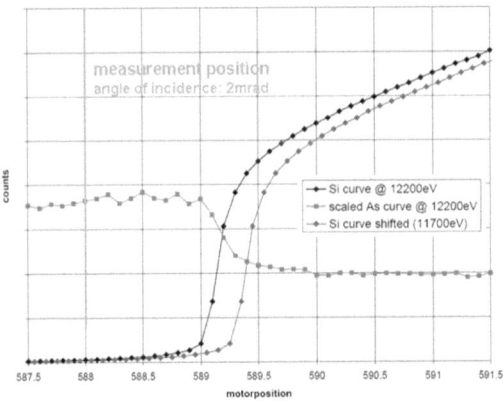

Figure 6.5: Anglescan of a 1 ng As droplet sample (residue on surface) on a silicon reflector. The change of the critical angle for Si for an energy scan from 11700eV to 12200eV is 0.11 mrad

At least two specimen have been analyzed for all six different types of samples (As(V) and As(III) standards, As(V) and As(III) containing nutrient solutions and xylem saps collected from plants grown in As(V) and As(III) containing nutrient solutions).

Simultaneously, the absorption by an elemental gold foil was recorded in transmission mode. The first inflection point (i.e. the first maximum of the derivative spectrum) of the Au metal foil scan was assumed to be 11918 eV (Au-L3 edge).

For quantification single fluorescence spectra recorded at 12200eV (figure 6.6) have been evaluated using the QXAS (Quantitative X-ray Analysis System) software package [116]. The arsenic concentrations in the xylem sap samples were calculated from sensitivities obtained by the measurements of the standard samples.

Absorption spectra have been analyzed within ATHENA which is included in the IFEFFIT program package for XAFS analysis [117-119]. The background removal of the As K-edge profiles was done by the implemented AUTOBK algorithm and normalization was performed by edge step normalization [119] using a pre-edge region ranging from -150 to -30 eV. For each scan, the energy scale was corrected with respect to the Au-L3 edge. Multiple scans of the same sample have been merged by calculating the average and standard deviation at each point in the set and scans of the xylem sap samples were smoothed by the removal of spurious points. A linear combination analysis of the K-Edge profiles was done with the fitting method of ATHENA. Linear combinations of the near-edge spectra for the standard solutions were fitted to those of the xylem sap samples.

6.3 Results and discussion

Quantification of the xylem sap samples revealed arsenic concentrations in the range of 30 to 50 ng/mL corresponding to an amount of 0.6 – 1 ng for the samples where 20µl have been pipetted. A representative spectrum obtained from a 20µl xylem sap sample is given in figure 6.6.

Detection limits for arsenic in xylem sap were determined by extrapolation for a 1000 s measuring time and found to be in the 0.2 ng/mL range. These results are in good agreement with the ones obtained with FIA which showed total arsenic concentrations of 29.5±1 and 45.7±2.6 ng/mL in xylem saps of plants treated with As(V) and As(III) respectively [154].

Multiple scans of each sample have been analyzed concerning the reproducibility of the measurements and to check if any alteration of the oxidation state occurred during the measurement. Figure 6.7 shows two examples of four repetitive XANES scans. In case of the nutrient solution the total measuring time was 100 minutes (25 minutes each), for the standard sample 220 minutes (55 minutes each). It is obvious from the plots that no edge shifts or changes in the oscillatory part of the spectra appeared during the measuring time. Therefore it can be concluded that the chemical state of the sample remained unchanged during the measurements.

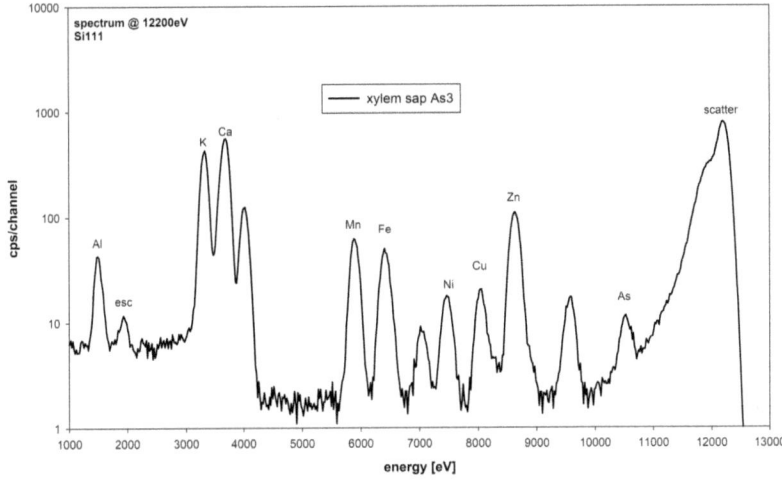

Figure 6.6: Fluorescence spectrum of xylem sap recorded at 12200eV

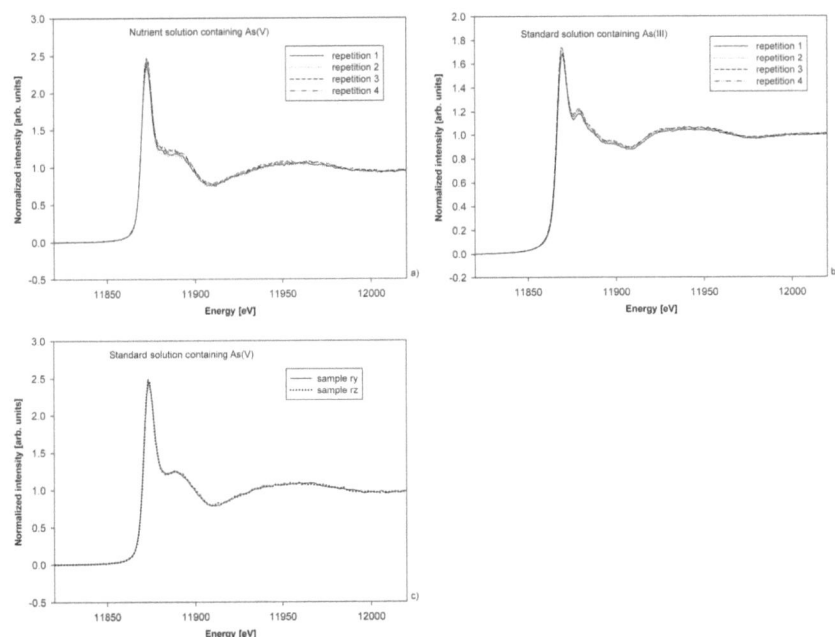

Figure 6.7: Repetitive XANES scans of two samples: (a) 10 µl of nutrient solution containing 750 ng/ml As(V) and (b) 20 µl of 10 µg/ml As(III) standard solution. Figure 2(c)) shows two scans of samples applied to different Quartz reflectors using the same standard solution (1 µl of 10 ppm As(V))

Normalized As K-edge profiles for xylem sap samples, nutrient solution samples and reference As standard samples are shown in figure 6.8. The spectra are displaced vertically for clarity. The vertical dotted line indicates the energy of the Au-L3 edge used for energy calibration. The vertical solid line marks the white-line (strongest absorption peak) of the As(III) standard spectrum to visualize the edge shifts. The two xylem profiles labeled 'xylem sap (As(III))' and 'xylem sap (As(V))' refer to samples collected from plants treated with nutrient solutions containing As(III) and As(V), respectively. The plot shows that the spectra of these two types of xylem samples have the same edge position which furthermore coincides with the edge position of the As(III) standard. These results indicate that As(III) is the predominant form in xylem saps although plants have been grown in nutrient solutions containing different arsenic species. The XANES spectrum of the As(III) nutrient solution collected 48 hours after the start of the arsenic treatment shows an energy shift towards the As(V) edge position. This indicates a partial oxidation of the As(III) to As(V) during this time period.

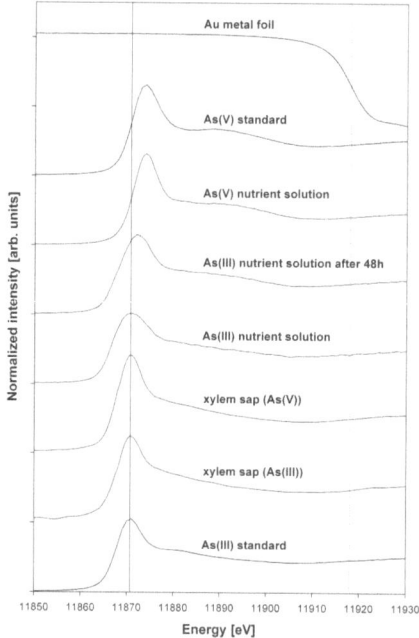

Figure 6.8: Normalized arsenic K-edge XANES spectra for the xylem sap, the nutrient solution and the As reference samples

All XANES spectra of nutrient solutions and xylem saps have been fitted with linear combinations of the spectra of the As(III) and As(V) standards. Table 6.1 shows the results of the processed fits and the quality-of-fit parameters R, chi square and reduced chi square. These parameters are defined by:

$$R = \frac{\sum (data - fit)^2}{\sum (data)^2} \qquad (6.1)$$

$$\chi^2 = \sum \frac{(data - fit)^2}{data} \qquad (6.2)$$

$$reduced \chi^2 = \frac{1}{n-m} \chi^2 \qquad (6.3)$$

with n-m = data points - variables

The fitting range was set from -20 to +50 eV relative to the edge and each fit included 96 data points and 1 variable (n-m = 95). The same parameters were used for fitting the spectra of nutrient solution samples.

Sample	As(III) [%]	As(V) [%]	R-factor	χ^2	reduced χ^2
xylem sap (As(III))	88 ± 3	12 ± 3	0.0155	1.09	0.0115
xylem sap (As(V))	83 ± 3	17 ± 3	0.0143	1.06	0.0112
As(III) nutrient solution	100 ± 3	0 ± 3	0.0103	0.68	0.0072
As(III) nutrient solution after 48h	71 ± 3	29 ± 3	0.0080	0.60	0.0063
As(V) nutrient solution	2 ± 2	98 ± 2	0.0065	0.63	0.0066

Table 6.1: Results of best linear combination fits for spectra of standard samples to those for xylem sap and nutrient solution samples

To double check the uncertainty of the fitting results, repetitive scans of the same sample haven been fitted individually to determine the influence of the measurement statistics. The uncertainty of the As(V)/As(III) ratio determination was found to be 2% for the fitting of the spectra of nutrient solutions and 5% for the fitting of xylem saps.

The results give quantitative information about the findings discussed qualitatively in figure 6.8. More than 80% of the arsenic in xylem sap was found to be As(III) independent of the arsenic treatment. Mihucz et al. [154] reported 86% As(III) in xylem saps which is in good agreement with these results. After 48 hours <30% of the As(III) in nutrient solutions was oxidized to As(V).

Chapter 6   Analysis of Arsenic in Cucumber Xylem Sap

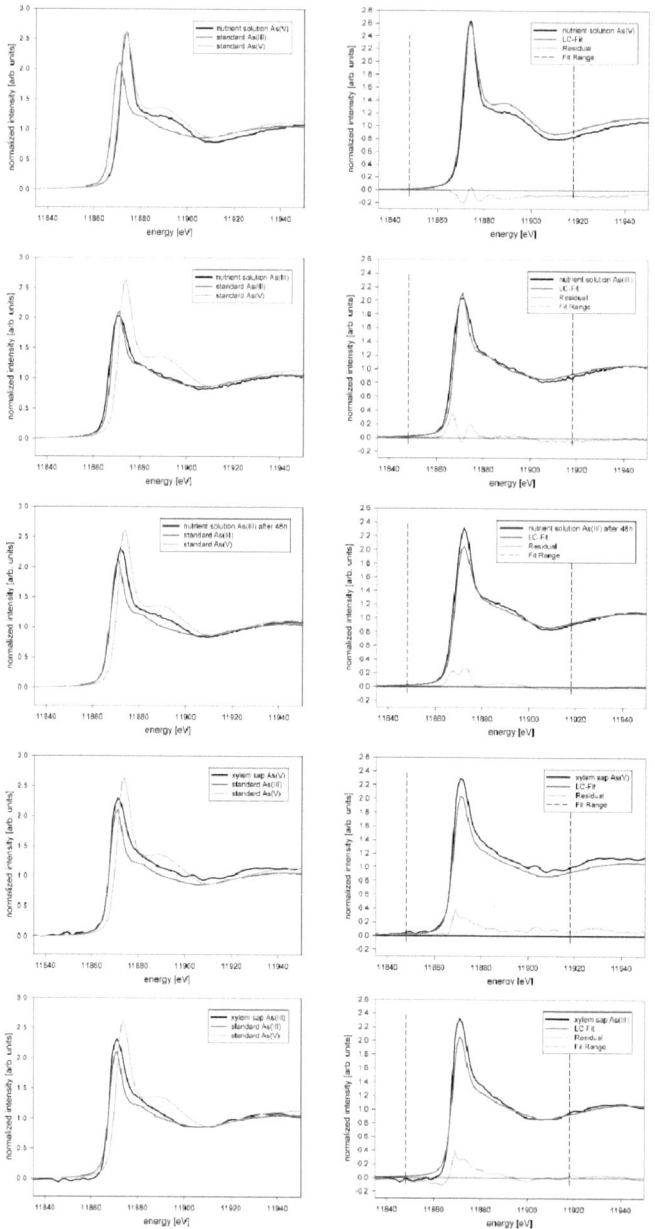

Figure 6.9: Left: XANES scans of fitted samples in comparison with the spectra of the arsenic standards. Right: Results of the best linear combination fits

During the measurements of the higher concentrated standard samples a damping of the white-line was observed. Figure 6.10 shows this effect on the basis of two XANES scans recorded for two As(V) standard samples made from the same standard solution containing 10μg/mL As(V). Different volumes of 1μl and 4x5μl have been pipetted on the reflectors for total amounts of 10ng and 200ng arsenic respectively.

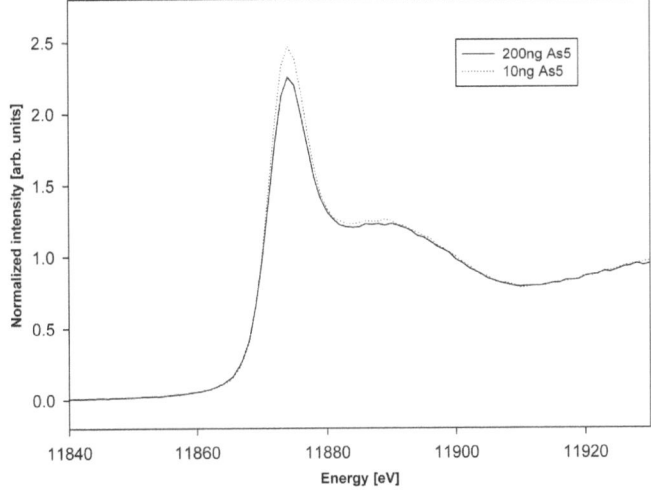

Figure 6.10: Arsenic K-edge XANES spectra for different total amounts of arsenate

To estimate the influence of this damping for the LC analysis samples have been fitted using the 10ng and the 200ng As(V) standard spectra. The differences in the determination of the As(V)/As(III) ratio have been found to be <3% for xylem sap samples and in the range of 1% for nutrient solutions. Both values are smaller than the calculated uncertainties of the LC analysis due to measurement statistics.

The damping of the white line could be explained by self-absorption effects due to the TXRF geometry. In this geometry the path length of the incident beam in the droplet is longer than in other geometries and therefore its absorption in the sample cannot be ignored for larger amounts of concentrated samples. The self-absorption effect concerning surface analysis with X-ray Absorption Fine Structure measurements in grazing incidence geometry has been studied by various authors [100, 105, 106]. However, the investigations did not consider droplet sample geometries. Therefore this topic needed further investigation and will be discussed in chapter 7.

6.4 Summary

It can be concluded that SR-TXRF offers good sensitivity for XANES speciation of chemical elements present in droplet samples at trace element levels. It could be demonstrated that a speciation of As is possible down to the 30 ng/mL level with this method. Repetitive measurements showed high reproducibility and no alteration of the oxidation state of the samples during the measurements. Due to the grazing incidence geometry, self-absorption effects for droplet samples with high concentrations have to be considered and will be discussed in chapter 7. This may lead to a further improvement of the analysis of TXRF-XANES spectra.

The presented data shows that cucumber plants treated with arsenate in concentrations of 150 ng/mL convert As(V) to As(III). A quantification of this effect reveals almost no difference in the ratio of As(V) to As(III) in the xylem sap of plants treated with nutrient solutions containing these two arsenic species. The presence of As(V) in the xylem sap of plants treated with As(III) containing nutrient solution suggests a partial oxidation of As(III) to As(V) in the nutrient solution before uptake. This suggestion could be assured as an analysis of the As(III) containing nutrient solutions revealed a partial oxidation of As(III) to As(V) (<30% after 48h).

All results concerning arsenic speciation in xylem saps and nutrient solutions are in good agreement with those obtained by HPLC-HR-ICP-MS [154]. This indicates the competitive capability of SR-TXRF XANES for trace element speciation.

Chapter 7

Self Absorption Effects in TXRF-XANES analysis

7.1 Introduction

The previous chapters showed that Total reflection X-Ray Fluorescence (TXRF) analysis in combination with X-ray Absorption Near Edge Structures (XANES) analysis is a powerful method to perform chemical speciation studies at trace element levels. However, when measuring samples with higher concentrations and in particular standards, damping of the oscillations is observed (see chapter 6). It is assumed that this is due to a self-absorption effect which smears the spectral shape and damps the fine structure. The term "self absorption" here means the absorption along the path of the incident beam. In this chapter the influence of self absorption effects on TXRF-XANES measurements will be investigated by comparing measurements with theoretical calculations.

7.1.1 The self-absorption effect

Fluorescence acquisition of X-ray Absorption Spectroscopy (XAS) spectra of concentrated samples suffers from self absorption which causes damping and also broadening of the oscillations. Some authors have performed quantitative speciation through analysis of XANES spectra by fitting them with analytical functions [166, 167]. Another approach deals with the corrections of the measured spectra to account for self absorption. Many authors have investigated self absorption effects in XAS using fluorescence acquisition depending on the angle of incidence and detection and have proposed correction models [105, 106]. Due to irregular sample shape and the very shallow angle of incidence these models are not applicable to TXRF. The extreme grazing incidence geometry used by TXRF enhances these self-absorption effects due to the extended path length of the incident beam in the droplet. This path length is equivalent to the penetration depth of the incident beam and is therefore energy dependent. As the energy changes during a XANES scan the size of volume where the fluorescence photons originate from is varying. Higher absorption means smaller excited volume and therefore less fluorescence intensity (and vice versa). Consequently this leads to a damping of the oscillations of the absorption coefficient above the absorption edge (figure 7.1). The phenomenon is relevant at higher concentrations, such as would be used in the measurements of standards to have a good counting statistic in relatively short measurement time. Any absorption of the fluorescence radiation within the sample can be neglected because it remains constant during an energy scan.

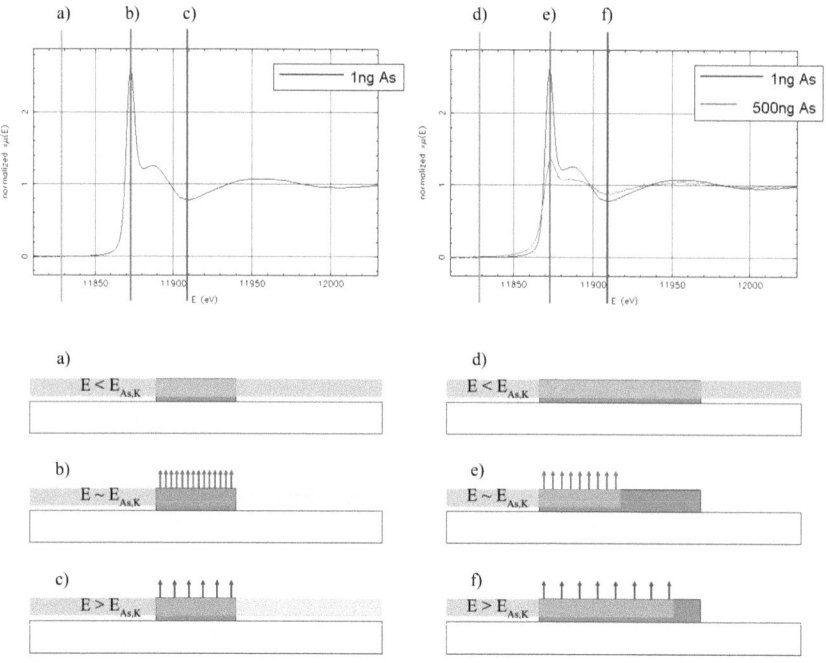

Figure 7.1: Schematic representation of the self-absorption effect. The incident beam (orange) is drawn parallel to the reflector's surface (blue) to account for the extreme glancing incidence geometry. Due to this geometry the beam has to pass through the lateral dimension of the sample (green). The pre-edge regions, the maxima ("white-line"), and the minima of two XANES scans for to samples with different total amounts of arsenic are indicated by a),b),c) and d),e),f) respectively. For cases a) and d) the energy of the incident beam E is below the energy of the K-absorption edge of arsenic ($E_{As,K}$). Almost no absorption takes place and no As-Kα fluorescence photons are emitted. As the energy is increased to the absorption edge (cases b) and e)) strong absorption and therefore maximum fluorescence intensity occurs. For the following it is important to note that the XANES scans show normalized and not total fluorescence intensities. That means the oscillations depend upon the relative intensities at different energies. If the lateral dimension of the sample becomes larger than the penetration depth of the incident beam (case e)) the normalized/relative fluorescence intensity is decreased. As the penetration depth increases with higher energy a larger volume of the sample is attenuated (case f)). This leads to a higher normalized fluorescence intensity compared to case c).

In general TXRF is known to allow for linear calibration typically using an internal standard for quantification [72, 75]. In fact for the fluorescence radiation collected by the detector, the sample is "thin" and differential absorption for photons with different energies can be ignored. Assuming the sample to be homogeneously distributed the loss of fluorescence signal due to absorption of the primary beam equally affects all elements and quantification by using an internal standard is not affected by the phenomenon.

7.1.2 Aims

The investigations in chapter 6 and other studies [99] showed a significant damping of the XANES spectra of standard samples. This effect may be correlated to self absorption effects. It is therefore of interest to investigate the absorption effect depending on:
- the total mass deposited
- the three-dimensional shape of the sample
- the density of the element investigated

A simple simulation model will be presented which represents a simple approach for an *a priori* evaluation of the self-absorption in TXRF X-Ray absorption analyses. The consequences for Extended X-ray Absorption Fine Structure (EXAFS) and XANES measurements under grazing incidence conditions will be discussed.

7.2 Experimental

Two series of samples have been measured. The first series consist of three samples on Plexiglas reflectors. Different total amounts of masses (1, 10 and 100 ng) of arsenic were applied onto the reflectors. The second series contained samples on Quartz reflectors with masses of 4, 9, 20, 72, 100 ng of arsenic and, additionally, one sample on a Plexiglas carrier containing 500 ng. This second series was analyzed with a confocal optical microscope to gain information about the 3-dimensional shape of the dried residues. For both series arsenic K-edge XANES measurements in fluorescence mode and grazing incidence geometry were performed. The results of these measurements were compared with the output of a simple Monte-Carlo simulation of absorption effects performed for each sample.

7.2.1 Sample preparation

A set of arsenic containing solutions had been prepared from a 1000+/-5 mg/L solution (MERCK CertiPUR, traceable to SRM from NIST, H_3AsO_4 in HNO_3, #1.19773.0100). Tri-distilled water (Atominstitut) was used for the dilution of the mother solution to obtain the following set of As concentrations: 1, 4, 10, 20, 40, 80, 100, 300, 500 mg/L. The respective volumes of mother solution and tri-distilled water were pipetted into a SARSTEDT flask, but the actual dilution factors were gained by the readings of a suited balance (SARTORIUS R300S).

1µL of each of these solutions was pipetted onto a Quartz or Plexiglas reflector (both 30mm diameter) respectively. Prior to the pipetting a blank measurement was done for each of these reflectors to assure that no As, Pb, etc. contamination would falsify the results. These measurements were performed with an ATOMIKA Extra II (LT: 100s, 50kV, 38mA). The beam spot size and the area inspected by the detector of the EXTRA II was in the range of various millimeters, therefore the examined area was much larger than the pipetted samples. The pipette (EPPENDORF research 0.1-2.5 µL) showed for this volume a maximum error of +/- 20%, consequently the pipetting step served only for sample positioning in the centre of the respective reflectors and the mass was determined by differential weighing:

Series 1 (Plexiglas reflectors.): 1.10, 10.02, 99.8 ng

Series 2 (Quartz reflectors.): 3.98, 8.99, 20.0, 39.8, 71.6, 100.2, 299.4, 503.5 ng

(The last sample of series 2 (500ng) was applied to a Plexiglas reflector)

The aqueous/acidic matrix was removed by drying the reflectors on a hot plate inside a flow hood in order to avoid contaminations. The Plexiglas reflector containing 500ng As was vacuum dried. An additional silicon reflector (30mm diameter) was prepared by applying 500ng Arsenic as well for comparison (see figure 7.5). This reflector was also dried in vacuum.

7.2.2 TXRF-XANES measurements

Arsenic K-Edge XANES measurements in fluorescence mode and grazing incidence geometry were carried out using the setup at the beamline L at the Hamburger Synchrotronstrahlungslabor (HASYLAB) at DESY [92, 95]. All measurements were performed in vacuum. A Si(111) double crystal monochromator was used for selecting the energy of the exciting beam from the continuous X-Ray spectrum emitted by the 1.2 Tesla bending magnet at beamline L. The primary beam was collimated to 200µm x 2000 µm (horizontal x vertical) by a cross-slit system. The incident X-ray intensity was monitored with an ionization chamber.

During the measurements the excitation energy was tuned in varying steps (5eV to 0.5eV) across the arsenic K-edge at 11867 eV. At each energy a fluorescence spectrum was recorded

Chapter 7 Self Absorption Effects in TXRF-XANES analysis

by a Silicon Drift Detector (SDD, VORTEX 50 mm², Radiant Detector Technologies) [114, 115]. The distance between SDD and sample carrier was 1mm [95].

The data acquisition time was increased for samples with lower total amounts of arsenic. This was done to achieve better statistics for the evaluation of the As-peak area. From these peak areas the XANES spectra were built and therefore a specific minimum count rate was desirable. For series 1 (Plexiglas reflectors) the acquisition time for each spectrum was set to 10 seconds for the 1ng sample and to 4 seconds for the 10 and 100ng sample. Each scan consisted of 305 spectra and the energy range was set from 11700 to 12300eV. For series 2 (Quartz reflectors) the acquisition times were set to 1 second for the samples containing masses of 300 and 500ng and to 2 seconds for the rest of the samples. The energy range was set from 11700 to 12500eV and 576 spectra were recorded for each scan of this series. Simultaneously, the absorption by an elemental gold foil was recorded in transmission mode for each scan of both series. The first inflection point (i.e. the first maximum of the derivative spectrum) of the Au metal foil scan was assumed to be 11918 eV (Au-L3 edge).

To check if any contaminations occurred due to sample handling single spectra of the XANES scans have been checked for contaminations. As example two fluorescence spectra of XANES scans (for the 40ng and 100ng samples on Quartz reflectors) are shown in figure 7.2:

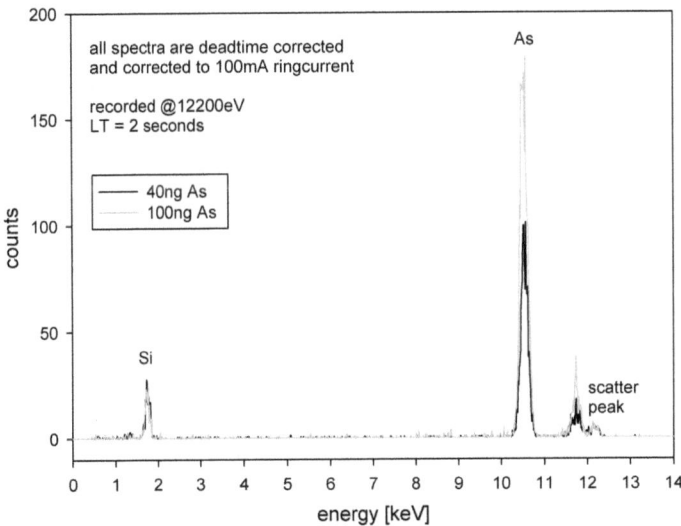

Figure 7.2: Single fluorescence spectra of two XANES scans recorded for the 40 ng and 100 ng samples on Quartz reflectors. It can be seen that no contamination of the samples occurred due to sample handling

105

As the critical angle of total reflection changes slightly (<0.2mrad for silicon) during the energy scan the incident angle of the primary X-ray beam was adjusted to 2mrad, which is far below the critical angle (~2.7 mrad at 11700eV). On that account it was assumed that the change of the critical angle during the XANES scans was unproblematic for the measurement of droplet samples (residues on surface).

Both measured and simulated absorption spectra have been analyzed with ATHENA which is included in the IFEFFIT program package for XAS analysis [117-119]. Using this software each scan was normalized and its energy scale was corrected with respect to the Au-L3 edge. Multiple scans of the same sample have been merged by calculating the average and standard deviation at each point in the set.

7.2.3 Confocal microscopy

Measurements to determine the shape of the samples of series 2 (Quartz reflectors and one Plexiglas reflector) have been performed utilizing a confocal white light microscope (NanoFocus μsurf® [168]). Additionally the silicon reflector (500ng As) was analyzed for comparison. The analyses were done by the Austrian Center of Competence for Tribology (AC2T). The measuring field was ~ 1450×1400 μm with a lateral resolution of ~ 1.5×1.5 μm and 50 nm in height. Due to the measurement set up it was necessary to perform a plane correction and flattening of the data:

The raw confocal microscope images were first leveled using a polynomial fit of grade 1, then the zero point of the z axis was set to the maximum of the histogram of the heights. The procedure was carried out using the software package SPIP 4.2.6.0 [169]. Additionally the data was treated using a threshold filter. This was done mainly because the amount of data had to be reduced to obtain reasonable simulation times. The threshold filter removed each data point with a height smaller than 25% of the maximum height of the sample. The result was a simplified sample with respect to the real sample's shape – speckles of questionable origin (traces of the sample, measurement artifacts or contaminations) which were found on the reflectors surface have been removed. The result of measurements and data treatment was a matrix representing the three-dimensional distribution of the sample on the reflector's surface (figures 7.3 and 7.4). This information was used for the simulation of absorption effects during a TXRF-XANES scan considering the sample's shape (see section 7.3).

Chapter 7   Self Absorption Effects in TXRF-XANES analysis

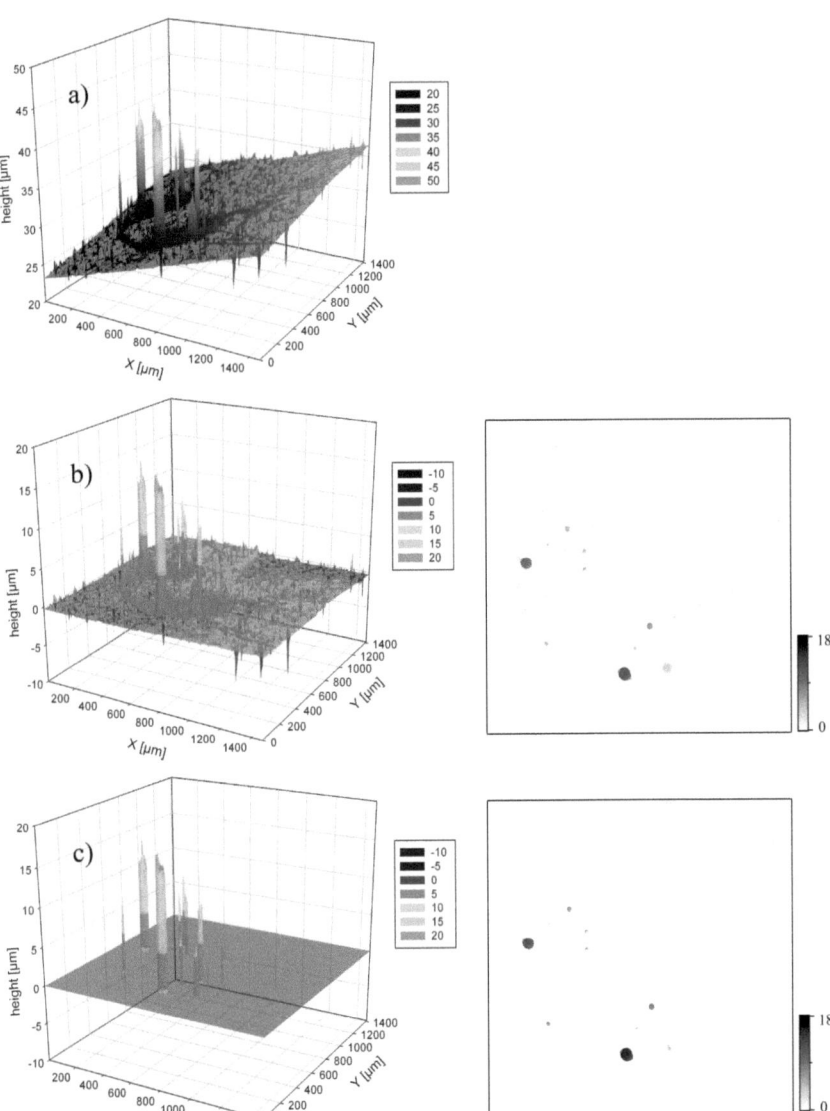

Figure 7.3: Correction procedure performed with the data obtained from the confocal microscope. As an example the data of the 100ng As sample is shown: a) the uncorrected data, b) the data after plane correction and flattening and c) after additional threshold filtering. The scale of the Z-axis (height) in the 3D illustration was enlarged for clarity. The right part of the figure shows the top view of the whole measuring field (~1450 μm ×1400 μm).

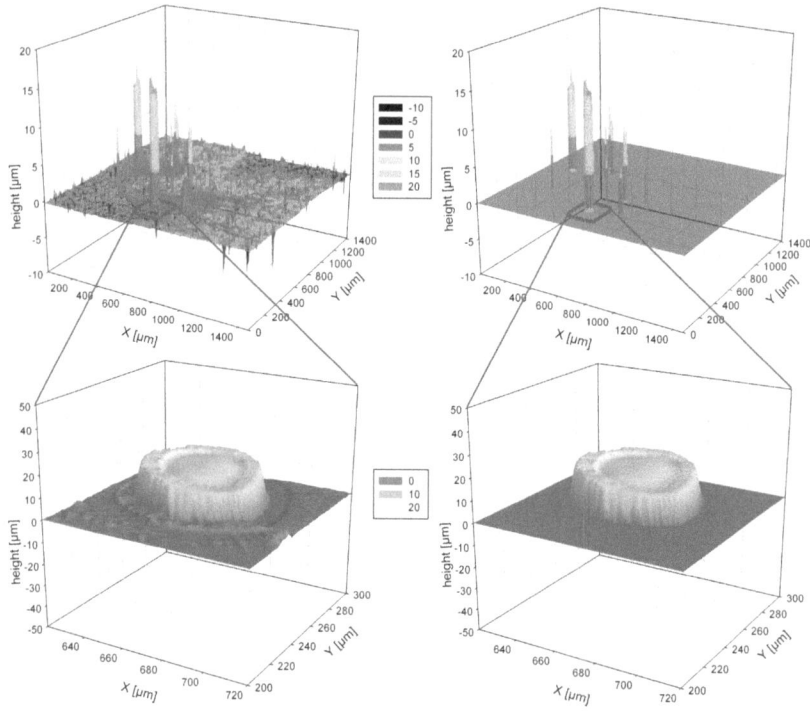

Figure 7.4: Zoom of the largest droplet of the 100ng As sample before (left) and after (right) threshold filtering. Unlike the upper part of the figure the scale of the z-axis (height) was not enlarged in the zoomed illustration to show the real relation between height and lateral dimensions of the droplet.

7.3 Development of a simple Monte-Carlo simulation

To simulate the observed damping of the XANES oscillations (e.g. figure 7.7) for samples with higher mass a simple Monte-Carlo simulation was developed.

XAS data obtained for the sample with the smallest mass was taken as reference XANES scan for the Monte-Carlo simulation (i.e. 1ng for the samples on Plexiglas reflectors (series 1) and 4ng for the samples on Quartz reflectors (series 2)). On the basis of preliminary investigations [99, 103] which showed damping effects for higher masses (>10ng) it was assumed that these scans show no significant damping of the oscillations of the fine structure. Each reference scan was normalized within the ATHENA software package by edge step normalization [119]. With this procedure the pre-edge region of the scan was normalized to zero while the post-edge region was normalized to one. The total absorption cross section µ used for the

Chapter 7   *Self Absorption Effects in TXRF-XANES analysis*

simulation has been calculated from the tabulated values for the photoelectric absorption cross section τ published by Henke et al. [80] and the scattering cross sections (σ_{coh} and σ_{incoh}) published by Ebel et al. [170]:

$$\mu = \tau + \sigma_{coh} + \sigma_{incoh} \quad (7.1)$$

For the calculation the value of the density of As in the actual sample is required (this is influenced by hydration when crystallizing in the process of drying and the presence of other elements present in the solution). Thus the µ used is linked to the coefficient for elemental arsenic through the following relationship:

$$\mu = \mu_{elemental}(E) \times (B/\rho_{As\_elemental}) \quad (7.2)$$

$\rho_{As\_elemental}$ = 5.73 g/cm³ (as tabulated for elemental arsenic)
B ("corrected density") is the actual density of the As in the sample
Here it was assumed that all the absorption in the sample is given by the As (the standard used is H_3AsO_4 in HNO_3, and there might be some hydration in the crystallization, but there are no heavy, strongly absorbing elements - this was verified by checking the single spectra of a scan for contaminations; see figure 7.2).

The energy scale of the measured data was corrected by shifting the maximum of the first derivative of the XANES scan to the maximum of the first derivative of the theoretical τ function. The K edge contribution τ_K to the total photoabsorption coefficient τ can be calculated according to

$$\tau_K(E) = \tau(E) \times (S_K - 1) / (S_K) \quad (7.3)$$

where S_K is the edge jump ratio.

For the simulation the tabulated absorption coefficients of a free atom were modified by superimposing the fine structure of the absorption coefficient measured in the reference spectrum. First a modified τ_K coefficient was calculated as follows:

$$\tau_{K,calc}(E) = \begin{cases} f_{ex}(E) \cdot S_K & \ldots E < absorption\ edge \\ \tau_K(E) \cdot f_{ex}(E) & \ldots E > absorption\ edge \end{cases} \quad (7.4)$$

where $f_{ex}(E)$ represents the normalized values of the reference scan.
In the next step the tabulated values of τ have been modified as follows:

$$\tau_{calc}(E) = \begin{cases} \tau(E) + \tau_{K,calc}(E) & \ldots E < absorption\ edge \\ \tau(E) - \tau_K(E) + \tau_{K,calc}(E) & \ldots E > absorption\ edge \end{cases} \quad (7.5)$$

The total absorption cross section $\mu_{calc}(E)$ was determined according to (7.1) and furthermore used for all calculations within the Monte-Carlo simulation.

Starting from a uniformly distributed random number P, interpreted as absorption probability of a photon in the sample, an absorption length L_{abs} was generated utilizing Beer-Lambert's Law:

$$L_{abs}(E) = -\log_e(P) / \mu_{calc}(E) \tag{7.6}$$

where $P \; (= I/I0 \;) \in \;]0,1]$

The absorption length is the path length of a photon propagating in the sample until it gets absorbed. For each incident photon penetrating the sample, the length L_{inc} of its path through the sample was compared with its absorption length L_{abs}. If the absorption length was found to be smaller than the incident photon's path length L_{inc} ($L_{abs} < L_{inc}$) an arsenic K-alpha fluorescence photon was counted with the probability $\tau_{K,calc}(E) / \mu_{calc}$. To simulate a whole XANES scan a fixed number of incident photons was used for each energy. The simulated XANES spectra were built from the counted K-alpha fluorescence photons per energy value generated as described above. The software for the simulation was written in C++ using the TT800 random generator [171] with a period of 2^800 to generate the random number P. The other input parameters L_{inc} and B have been determined in different ways for the two series of samples.

For the samples which were investigated with the confocal microscope (series 2) the real sample's dimensions were known from these measurements. Therefore the volume of each sample could be easily calculated to determine the density B:

$$B = m_S / V_c \tag{7.7}$$

V_c: volume of the sample

m_S: Arsenic mass

For these samples not only one L_{inc} value was calculated but a set of paths with different lengths depending on the shape of the sample. Assuming a non-divergent beam parallel to the reflectors' surface, the length of each path became a function of the Y and Z coordinate of the origin of the exciting photon (i.e. position where the incident photon entered the sample, see figure 7.5).

Chapter 7   Self Absorption Effects in TXRF-XANES analysis

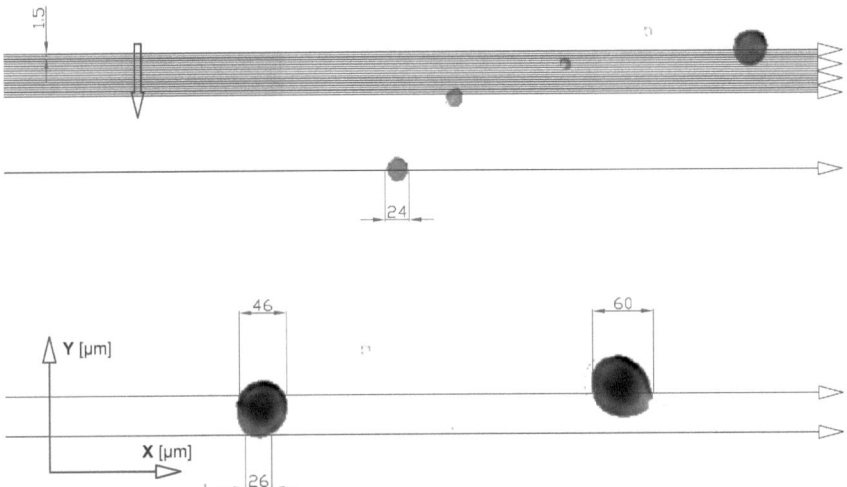

Figure 7.5: One zoomed region of the "72ng As(V)" sample of series 2. The dark spots are some of the dried residues of the pipetted sample. The arrows labeled 1 - 3 show three different possible paths of a photon propagating parallel to the sample carrier. It can be seen that for each of these three paths the photon covers different distances in the dried residues. These different distances (e.g. 106µm for path 2) are indicated in microns. To get a set of distances for all possible paths through the sample a path was calculated every 1.5µm (lateral resolution of the confocal microscope) along the Y-axis. This is indicated in the upper part of the figure (label 4). The calculation of the distances produced by each path was done for the entire Y-range of the measuring field of the confocal microscope.

The set of paths for each sample was calculated with respect to the lateral resolution of the confocal microscope (~1.5µm x 1.5µm). Thus one path through the sample was calculated every 1.5µm along the Y-axis as shown at the top of figure 7.5. This was done for the entire Y-range of the measuring field of the confocal microscope assuming that the whole sample was illuminated by the incident beam.

Considering not only the lateral but also the vertical dimension of the sample, this set of paths was calculated at different heights ranging from zero to the maximum height of the sample. For the simulation only paths with lengths L_{inc} larger than zero were taken into account.

During the simulation a fixed number of incident photons with one energy value were sent along each of these paths through the sample. So the decision if a fluorescence photon was produced or not had to be made N-times by the algorithm described above, where N is given by:

$$N = \text{number of energy points} \times \text{number of photons} \times \text{number of paths (length > 0)} \quad (7.8)$$

To keep the number of paths in a reasonable frame the influence of different step sizes along the Z-axis upon the simulation results was investigated and found to be negligible in the range of 100nm to 1000nm steps. Therefore the 1000nm step size was chosen for all simulations which resulted in much shorter simulation times.

For the samples which have not been measured with the confocal microscope (series 1) the parameters have been calculated assuming simple sample geometries. The residue of the droplet on the reflector surface was assumed to be a cylinder with diameter d and height z. With this simple geometry L_{inc} has been calculated as the mean path of a photon penetrating the cylinder parallel to the reflector's surface:

$$L_{inc} = d/4 \times \pi \qquad (7.9)$$

For the calculation of the diameter d the unknown parameters B ("corrected density") and sample height z have been estimated from the data of samples with similar mass of series 2 (i.e. the "9g As(V)" and the "100ng As(V)" samples; see table 7.1):

$$m_S/B = V_c = (d/2)^2 \times \pi \times z \qquad (7.10)$$

V_c: the volume of the cylinder

m_S: Arsenic mass

Even though these parameters (B, z) have been determined for different samples the simulations showed very good consistence with the measurements (see section 7.4.3).

In the following a short summary of the Monte-Carlo simulation parameters is given:

General assumptions:

- no beam divergence
- beam parallel to reflector-surface
- beam illuminates whole sample

Parameters for samples of series 1 (no confocal microscope measurements performed):

- simple sample geometry (cylinder, diameter d, height z)
- 100000 photons per simulation

Parameters for samples of series 2 (analyzed with confocal microscope):

- 1000nm step size in Z-direction (height) for calculations of paths through sample
- 1000 photons per path through sample

7.4 Results and discussion

7.4.1. Results of the measurements with the confocal microscope

Confocal microscopy images were collected for all the samples of series 2. Additionally the silicon reflector (500ng As) was analyzed for comparison. All results of measurements and data treatment are reported in figure 7.6. The figure displays the top view of the whole measuring field (~1450 µm ×1400 µm) for all samples. The information about the height of each sample is given by the corresponding red-intensity scale. The datasets presented in the figure were corrected using the procedures described in section 7.2.3.

In the following the most eye-catching characteristics of the results are discussed:
- All samples applied to Quartz reflectors dried forming droplet-like residues of different size. Except for the 4ng and 40ng sample the residues were aligned forming a ring.
- The sample applied to the silicon reflector (500ng) showed a similar behavior, but formed one large residue containing almost the whole mass.
- The sample applied to the Plexiglas reflector dried forming a ring containing much more but smaller residues.
- The 300ng sample was distributed over an area almost twice the size of the areas covered by the other samples.

The last point represents a serious problem as the beam width was set to 2000 µm for the TXRF-XANES analysis of the samples. The 300ng sample was therefore not fully illuminated by the beam.

A theory for the formation of a ring by solids dispersed in a drying droplet is described by Deegan et al. [172-174]. This model is very robust since it is independent of the nature of the solute. According to this theory an outward flow within then droplet transports the solute to the contact line. (The contact line is the border where the surface of the droplet contacts the surface of the carrier). This flow occurs when the contact line is pinned so that liquid which is removed by evaporation from the edge of the drop must be refilled by a flow of liquid from the inside. The reasons for the contact line pinning are irregularities of the substrate: surface roughness or chemical heterogeneity. Deegan et al. [174] reported that no ring was formed when the pinning was eliminated by drying the drop on smooth Teflon. In this case the drying drop contracted as it dried. An important point is that the contact line cannot be pinned permanently by the substrate. However, if solute particles are accumulated at the contact line the pinning is strengthened. This effect is called "self-pinning" [173]. An expansion of the theory of Deegan was given by Popov [175] taking into account the volume occupied by the solute particles.

Chapter 7   Self Absorption Effects in TXRF-XANES analysis

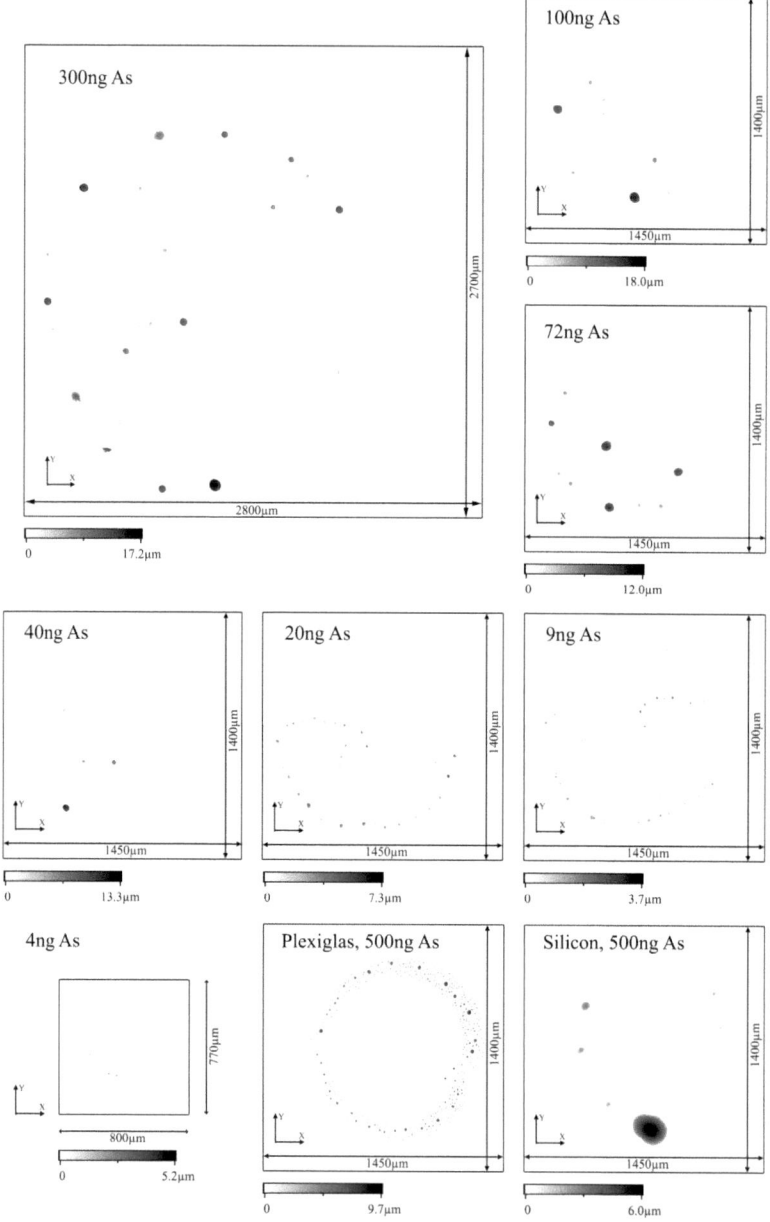

Figure 7.6: Results of all measurements with the confocal microscope. The data was corrected as described in section 7.2.3. It is striking to note that the shape of the sample on the Plexiglas and silicon reflector is totally different even though the sample preparation was identical.

Referring to the theory proposed by Deegan and Popov the formation of the residues could be explained as follows:
Some irregularity of the reflectors surface anchors the contact line of the deposited droplet at one or more points. As the liquid evaporates the solute particles are transported to these points and increase the pinning of the contact line. Eventually the primary droplet is disrupted to smaller droplets – one for each anchor point. Now these smaller droplets either undergo the same procedure again or contract at the anchor point forming residues like the one shown in figure 7.4. The occurrence of arc- and spiral-like alignment of the residues (samples 100, 20 and 9 ng) could be explained by repetitions of this procedure. The shapes of the residues on the Plexiglas and silicon reflector can be interpreted as antipodal extreme examples. On the one hand the surface of the silicon reflector has a very small roughness and is highly hygroscopic. Therefore the contact line could be pinned only at a few points and almost the whole primary droplet contracted at one point. On the other hand the Plexiglas reflector has a higher surface roughness and is less hygroscopic. Hence the whole contact line of the primary droplet is pinned at first. The solute particles are transported to the contact line and finally build a large number of small residues. Maybe this happens because they are themselves new anchor points.

7.4.2. TXRF-XANES measurements

The figures 7.7 and 7.8 show the results of the XANES measurements utilizing TXRF geometry which have been performed with sample series 1 and series 2 respectively. It can be clearly seen, that the damping of the white-line and post-edge oscillations can be correlated to the total mass of arsenic deposited.

As already mentioned in section 7.4.1 the 300ng sample was not fully illuminated by the beam because it was larger than the beam spot size. This is a possible reason why the damping of the oscillations of the corresponding XANES scan shown in figure 7.8 is smaller than expected. Another interesting fact is that the XANES of the 500ng samples on the Plexiglas and silicon reflector were almost identical although their shapes were totally different (see figure 7.6).

With the energy resolution used for the measurements an energy shift was observed for the 500ng and the 100ng samples (about 1.5eV and 0.5eV distance between the maxima of the first derivatives of sample 4ng and samples 500ng and 100ng respectively). The scan for the 1ng sample was used as reference scan for the Monte-Carlo simulations of series 1 and the scan of the 4ng sample for the simulations of series 2.

Chapter 7   Self Absorption Effects in TXRF-XANES analysis

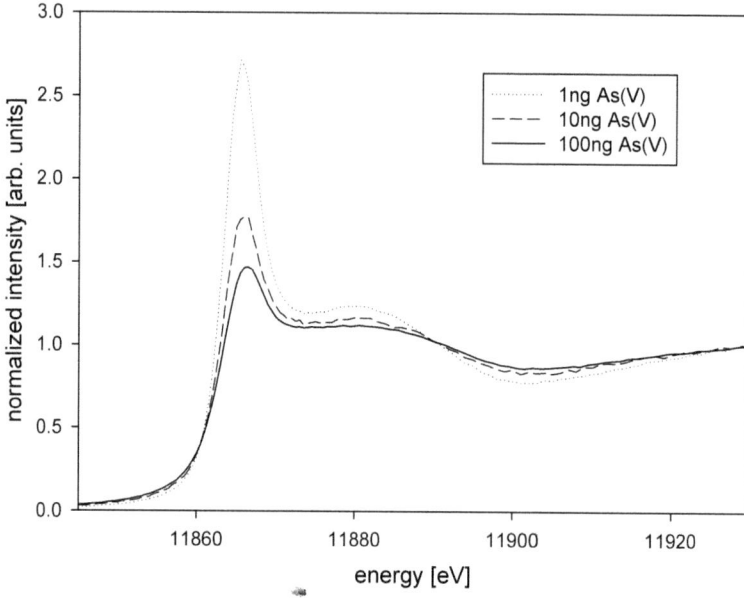

Figure 7.7: XANES scans of sample series 1 (1, 10, 100ng of As(V) on Plexiglas reflectors) showing the damping of the oscillations. The correlation between the absolute sample mass and the damping of the white line and the post-edge oscillations can be clearly seen.

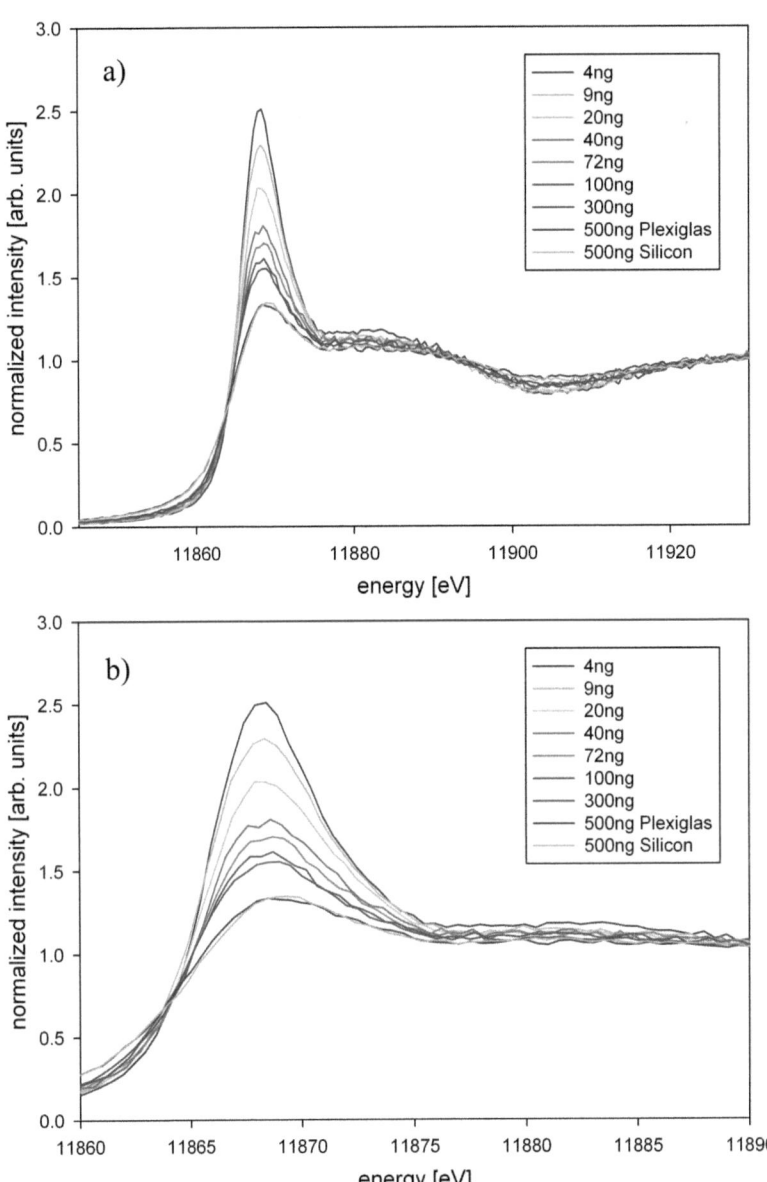

Figure 7.8: XANES scans of sample series 2 (4 - 500ng of As(V) on Quartz, Plexiglas and silicon reflectors) showing the damping of the oscillations. In b) the near edge region of the scans is enlarged. Like in figure 7 it is obvious that the absolute sample mass and the damping of the white line and the post-edge oscillations are correlated.

7.4.3. Results of the Monte-Carlo simulations

Figure 7.9 shows a comparison of the best simulations for series 2 and their corresponding measured scans. The simulations show good agreement with the measured scans. The damping and broadening of the white line as well as the damping of oscillations at higher energies could be reproduced sufficiently well. The simulations for the 40ng, 300ng and 500ng silicon samples produced no satisfactory results. This was no surprise concerning the 300ng sample due to the reasons mentioned in sections 7.4.1 and 7.4.2. The simulations performed for the 40ng and 500ng silicon samples showed too strong absorption effects. A possible reason for this could be the higher value of parameter B ("corrected density") calculated for these samples (see table 7.1).

For the simulation of the XANES scans related to the samples of series 1 which have not been investigated with the confocal microscope the parameters B ("corrected density") and z (cylinder height) had to be estimated as described in section 7.3. These parameters defined the dimensions of the cylinder which was the simple approximation of the samples' shape for the calculations. Even though the values of these parameters have been roughly estimated from the data of different samples (series 2) the simulations showed good agreement with the measurements (figure 7.10, table 7.1).

Simulation for sample (series 2):	maximum height z [µm]	"corrected density" B [g/cm³]		chi² *
9ng As(V)	3.7	1.22		0.1724
20ng As(V)	7.3	1.24		0.1062
40ng As(V)	13.3	1.92		0.4415
72ng As(V)	11.9	0.96		0.2831
100ng As(V)	18.0	1.25		0.2126
300ng As(V)	17.2	1.19		
500ng As(V) Plexiglas	9.7	2.69		0.1986
500ng As(V) Silicon	5.9	4.19		3.6081
Simulation for sample (series 1):	height of cylinder z [µm]	"corrected density" B [g/cm³]	Ø of cylinder d [µm]	
10ng As(V)	4.0	1.22	51.1	0.1899
100ng As(V)	15.0	1.25	82.4	0.4839

Table 7.1: Summary of simulation parameters
*: chi² of the simulation was calculated for the energy range 11800eV to 12000eV. The 300ng sample was not considered due to the reasons mentioned in sections 7.4.1 and 7.4.2

Table 7.1 shows the key parameters of the best simulations for all samples. To evaluate the quality of the simulations the goodness-of-fit parameter chi square was calculated for each simulation: chi^2 = Sum ([simulated value(E_i) − measured value(E_i)]^2 / measured value(E_i)).

For the calculation of chi square only the XANES region was considered as the interest focuses on the oscillations in this energy range (11800 − 12000eV).

The values of B for series 2 have been calculated from the data obtained from the measurements with the confocal microscope (i.e. the volume of the samples) and the known mass of the samples. The determined values are small in comparison with the tabulated value of the elemental density of arsenic (5.73 g/cm^3) due to hydration during the crystallization process and show a surprising consistency (except for the 40ng and both 500ng samples). With the exception of the 40ng, 300ng and both 500ng samples the maximum heights of the samples show an exponential trend. It is striking to note that the same samples (except the 500ng Plexiglas sample) produced no satisfactory results in the simulation. It seems that the "corrected density" B was the key parameter for the simulation of the self absorption effect (the 300ng sample showed a consistent B value but was not considered due to the reasons described above.)

Because of this consistence of B for the samples of series 2 this parameter was kept constant for the simulations done for series 1. The height of the cylinder z was also estimated from the values gained for sample series 2. It was then slightly varied to optimize the simulation with respect to the value of chi square.

Chapter 7   Self Absorption Effects in TXRF-XANES analysis

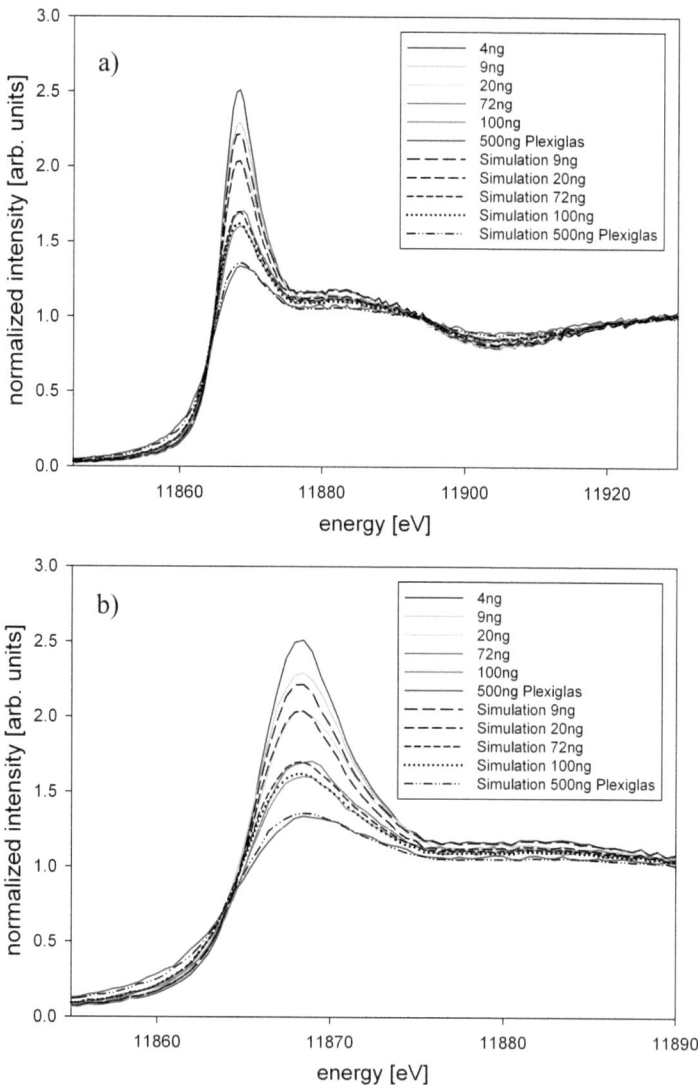

Figure 7.9: Measurements and best simulations for the samples of series 2. In b) the near edge region of the scans is enlarged

Chapter 7 Self Absorption Effects in TXRF-XANES analysis

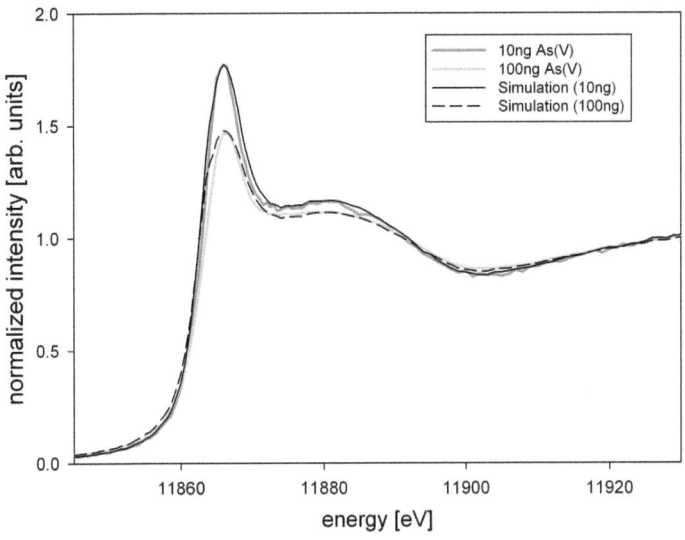

Figure 7.10: Measurements and simulations for the samples of series 1

7.5 Summary

The topic of the presented work was the influence of self-absorption effects on TXRF-XANES measurements. The effect of the sample shape as well as the one of different concentrated droplet samples was studied by comparison of calculations and measurements. XANES measurements of two sample series each with different total amounts of mass deposited on Quartz and Plexiglas reflectors were carried out under grazing incidence conditions. The results showed a linear correlation of the damping of the oscillations of the scans with the total mass of the samples. It was assumed that this attenuation originates from self absorption effects caused by the extreme grazing incidence geometry. To verify this hypothesis a simple Monte-Carlo algorithm was developed to simulate these effects. The data about the geometry of the samples required for the simulations was obtained by measurements with a confocal microscope. The simulations performed with this data showed good agreement with the measurements confirming the influence of sample mass and geometry on the damping of the oscillations. On the basis of the data obtained by the measurements with the confocal microscope samples with unknown shape have also been simulated. The results

Chapter 7   Self Absorption Effects in TXRF-XANES analysis

showed good correlation with the measurements as well. The key parameters of this study were the "corrected density" and the length of incidence beam in the sample. It seemed that the shape of the sample is of less importance than the "corrected density" which showed good agreement for (almost) all samples. However, the expected differences in the shape of dried residues on Plexiglas and Quartz reflectors due to different surface characteristics should be topic of further investigations. Another important point related to this topic is the influence of the sample preparation. The deposition of the aqueous sample and the drying process seems to be crucial for quantification in TXRF (especially) without using an internal standard. The best approach to overcome absorption problems seems to be an array of small spots produced with picoliter-pipettes or inkjet printers [138]. This topic should also be further investigated.

The presented results showed that performing an Extended X-ray Absorption Fine Structure (EXAFS) analysis under grazing incidence conditions for higher concentrated samples is very difficult. The damping of the oscillations would make a study of the EXAFS signal almost impossible. A direct correction of the measured scan is not possible because of the loss of information that the phenomenon brings about. For dilute samples on the other hand the measurement time has to be increased drastically to get reasonable counting statistics.

With TXRF acquisition for XANES used as a fingerprint method the investigated self-absorption effect is not dramatic. The energy position of the absorption edge is slightly affected for very high concentrations. This effect does not hinder quantitative evaluations, especially if analysis is carried out by fit of the XANES spectra with analytical functions. However, for a quantitative analysis performed by fitting of scans of unknown samples with those of known reference samples (linear combination method) it would be desirable to have undamped references. Therefore a compromise between counting statistics, measurement time and absorption effects has to be found for the measurement of standard samples.

The presented work proposes a rather simple way to study *a priori* the absorption effects that will show up in TXRF XANES and allow the scientist to prepare the sample according to needs (measuring time, acceptable self absorption) for the actual experiment. Moreover the method could be extended to allow the *a posteriori* correction for the self absorption of higher concentration standards.

Chapter 8

Comparison of Grazing Exit and TXRF geometry for XANES analysis

8.1. Introduction

The previous chapters were dealing with total reflection X-ray fluorescence (TXRF) analysis and absorption spectroscopy accomplished under total reflection conditions. A measurement setup utilizing an angle of incidence in the range of the angle of total reflection (± a few mrad) is called "grazing incidence" or "glancing incidence" (GI) setup. Therefore the total reflection geometry is a special case of grazing incidence measurements. The angle scans presented in chapters 4 and 6 (figures 4.4 and 6.5) show a typical angle range of grazing incidence measurements. These scans have been performed with the intention to adjust total reflection conditions. In this chapter the inverse GI setup and its applicability to XANES analysis will be presented. Results of measurements performed with the samples described in chapter 7 will be shown.

8.1.1 Grazing Exit (GE) geometry:

In contrast to grazing incidence X-ray fluorescence (GI-XRF), it is also possible to excite under normal incidence and detect the fluorescence radiation under glancing angle. This method is called "Grazing Exit" [90, 176-178], "Grazing Emission" (GE-XRF) [87-89, 107] or "Glancing-takeoff" XRF [179, 180] and was explained in section 3.3.3.3. It was shown that a GE experiment provides the same information as the GI experiment according to the optical reciprocity theorem [64]. As already mentioned in section 3.3 TXRF operates with the incident beam impinging below the critical angle of total reflection on the surface of a flat polished surface of reflector. The interference between incident and reflected beam causes in case of microcrystalline samples an intensity increase of the fluorescence signal by a factor (1+R) where R is the reflectivity numerically close to 1. For the GE geometry the interference is not between primary and reflected beam but among the superposition interference of the fluorescent waves emitted from the sample and observed under the critical angle of total reflection. As a result of the setup a much smaller solid angle is seen by the detector leading to the drawback of a lower sensitivity compared to TXRF. On the other hand it was suggested [105, 106] that a normal incidence-grazing-exit geometry would not suffer from self-absorption effects in XAFS analysis due to the minimized path length of the incident beam through the sample. Furthermore the critical angle of total reflection does not change during the energy scan of a XAFS measurement because the interference in GE is not between exciting and reflected wave field. Another advantage of the GE setup is the possibility to use a focused micro-beam of (synchrotron) radiation for excitation. This allows spatially resolved investigations of the sample with the advantage of the surface sensitivity of TXRF. A combination with XAFS analysis is also possible [91].

8.1.2 Aims

In the previous chapters the influence of the self absorption effect on the damping of the oscillations of XANES spectra has been discussed. To investigate if the normal-incidence grazing-exit geometry does not suffer from this phenomenon GE experiments have been performed using the samples which have already been analysed in TXRF geometry (chapter 7). The performance of the GE setup should be investigated with respect to the following points:

- Detection limits (in comparison to TXRF)
- Mapping capability
- XANES measurements (in comparison to TXRF)

8.2 Experimental

A GE-XRF experiment was performed at HASYLAB beamline L using the newly designed equipment from the Atominstitut Vienna X-Ray group. The setup utilizes the fact that Silicon Drift Detectors are lightweight and can therefore be easily moved by translation stages. This makes it possible to fix the sample position and keep the angle of incident (90°) constant during an angle scan. In figure 8.1 the setup and the fundamental arrangement is shown. The setup was designed with the axis of rotation of the detector exactly in the plane of the reflector. A sample holder was constructed to mount this reflector in a precise geometry. Easy sample changing by a motorized translation stage was available. The samples were dried spots prepared after pipetting a few microliters of a solution in the center of the reflector. A Peltier-cooled Si drift detector with 50mm² active area [114, 115] was used to collect the emitted fluorescence radiation under grazing angles. The detectors effective area of collection was delimited by the slit width (40μm or 200μm) of the defining diaphragm in front of the detector and was therefore roughly 40μm x 8mm or 200μm x 8mm.

The angle and the angular divergence of the fluorescence beam were defined by the slit width of the diaphragm and the distance between detector entrance slit and sample (40mm). These dimensions gave a theoretical angular resolution of 1mrad (40μm slit width) or 5mrad (200μm slit width). The dimensions of the incident beam of SR were set to 1600μm x 2500μm (horizontal x vertical) by a cross-slit system to assure that the whole sample was illuminated. In order to achieve measurements with higher lateral resolution as well a polycapillary half lens was used to produce a beam spot of 40μm in diameter.

All measurements were performed in air. Therefore the silicon signal of the reflectors could not be detected because it was absorbed on the relatively long distance (40mm) between sample and detector. A Si(111) double crystal monochromator was used for selecting the energy of the exciting beam from the continuous X-Ray spectrum emitted by the 1.2 Tesla bending magnet at beamline L. The incident X-ray intensity was monitored with an ionization chamber.

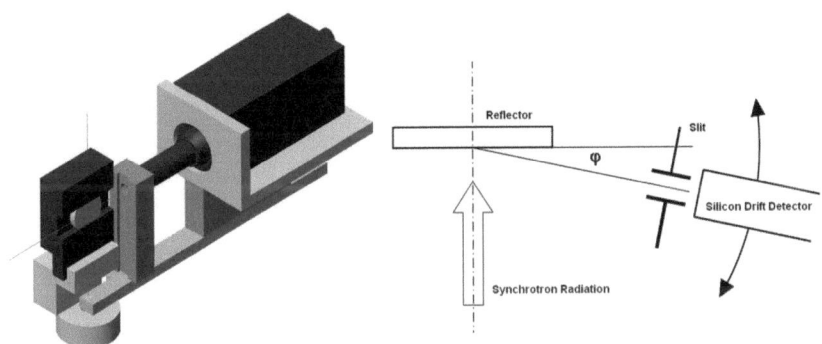

Figure 8.1: Left: View of the experimental setup of the GE measurements. Right: Sketch of the fundamental arrangement.

8.2.1 Samples

Three samples with different total amounts of arsenic masses on Quartz reflectors (20ng, 100ng) and a Silicon reflector (500ng) stemming from the sample series described in section 7.2 have been investigated. Additionally a Germanium reflector was used to perform angle scans (the silicon signal could not be used due to the absorption in air). This was done to compare the experimental results of the angular scans with theory.

8.2.2 Grazing exit measurements

Four kinds of measurements have been performed using the GE setup:
1. Angular scans of Arsenic samples and Germanium reflector using the both the unfocused and the focused beam (40μm beam spot)
2. Single TXRF spectra (LT 100s) of the samples (unfocused beam; 200μm slit width)
3. Area scans of the samples using the polycapillary half lens (40μm beam spot) and the 40μm entrance slit in front of the detector.
4. XANES measurements of the Arsenic samples (unfocused beam; 200μm slit width; two samples also with 40μm beam spot and 40μm entrance slit)

Angle dependent measurements were carried out by rotating the detector around the Arsenic droplet sample in the center of the reflector. The results of the angular scans were used to adjust an exit angle below the critical angle of total reflection in order to record single spectra, perform XANES measurements and accomplish the area scans.

For quantification the single fluorescence spectra recorded at 12200eV (figure 8.4) have been evaluated using the QXAS (Quantitative X-ray Analysis System) software package [116]. The Limits of Detection (LD) have been determined according to:

$$LD = \frac{3 \cdot \sqrt{N_B}}{N_N} \cdot m_{sample} \qquad (8.1)$$

where N_N and N_B are the netto- and background intensities of the signal and m_{sample} is the sample mass.

Table 8.1 shows the parameters of the XANES measurements. The excitation energy was tuned in varying steps (5eV to 0.5eV) across the arsenic K-edge at 11867 eV and a fluorescence spectrum was recorded for 10 seconds at each energy.

To calibrate the energy of the exciting radiation the absorption by an elemental gold foil was recorded in transmission mode. The first inflection point (i.e. the first maximum of the derivative spectrum) of the Au metal foil scan was assumed to be 11918 eV (Au-L3 edge).

Measured absorption spectra have been analyzed with ATHENA which is included in the IFEFFIT program package for XAS analysis [117-119]. Using this software each scan was normalized and its energy scale was corrected with respect to the Au-L3 edge. Repetitive scans of the same sample have been merged by calculating the average and standard deviation at each point in the set.

Energy regions of exciting radiation	Step width [eV] in energy region	Sample time [s]
11700	5	10
11750	2	10
11800	1	10
11870	0.5	10
11970	1	10
12100	2	10
12300	5	10
12500		10

Table 8.1: Energy-scan regions for XANES measurements with corresponding step widths and sample times

8.3 Results and discussion

8.3.1 Angular scans

Figure 8.2 shows the results of the angle dependent measurements of the bulk Germanium reflector in comparison with the theoretical curves. The data obtained for the angular scan performed with the focused beam (40µm spot size) and the 40 µm slit in front of the detector is displayed in figure 8.2a. The second measurement presented in figure 8.2b was accomplished using the unfocused beam (1600µm x 2500µm, horizontal x vertical) and the 40µm diaphragm. The critical angle of total reflection in GE geometry for Germanium was calculated to 4.33 mrad. To estimate the divergence of the measurements the theoretical curve was convolved with a Gaussian function with peak area one. The fit parameter was the full width at half maximum (FWHM) of the Gauss peak. The experimental data and the theoretical curve showed good agreement.

The divergence was expected to be 1 mrad for the measurement with the 40µm slit. However the divergence determined by the fitting was estimated to be 0.34 mrad for the experiment performed with the focused beam and 0.39 mrad for the angular scan using the unfocused beam. This low divergence can be explained with a misalignment of the slit relative to the reflectors surface. The mentioned effect was expected and was not corrected in order to have a higher angular resolution.

Chapter 8   Comparison of Grazing Exit and TXRF geometry for XANES analysis

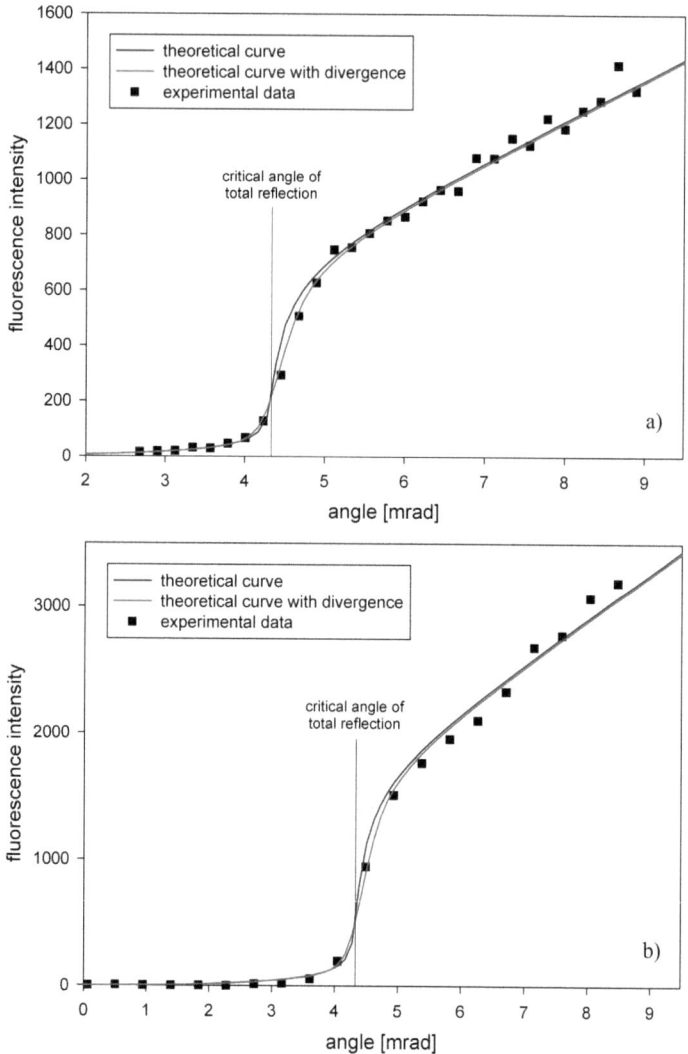

Figure 8.2: Angular dependence of the fluorescence intensity of the bulk Germanium sample. In a) the results of the measurements using the focused beam (40μm spot size) are shown. The data obtained using the unfocused beam is shown in b). Both measurements were performed using the 40 μm slit in front of the detector. The vertical line indicates the critical angle of total reflection at 4.33 mrad.

In figure 8.3 the fluorescence signals of the 100ng and 500ng As samples are displayed showing the typical shape with double intensity at angles below the critical angle when compared to larger angles. In GE geometry this effect is caused by the missing contribution due to surface reflections of the fluorescence signal in this region (see section 3.3.3.3). At

2.96 mrad the theoretically determined critical angle of total reflection in GE geometry for Arsenic is indicated. It can be seen that the angular divergence is larger for the scan using the 200μm diaphragm but is sufficient to determine an angle below the critical angle of total reflection for the further measurements (single TXRF spectra, area scans and XANES measurements).

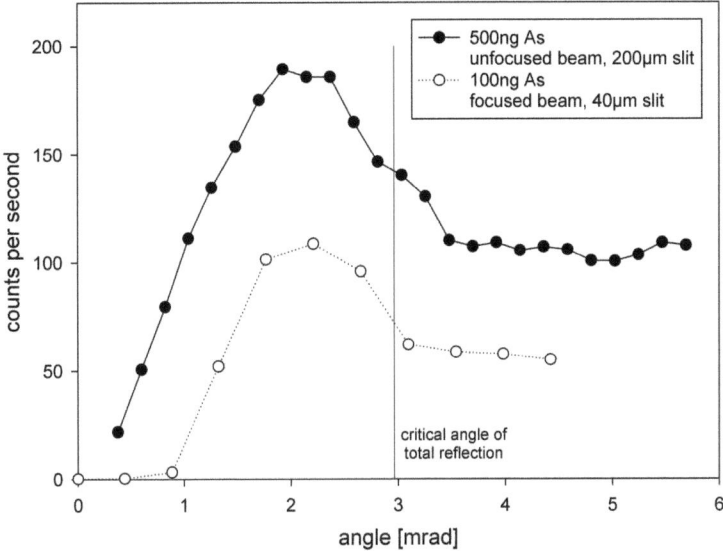

Figure 8.3: Angular dependence of the fluorescence intensity of the 100ng and 500ng Arsenic samples. The vertical line indicates the critical angle of total reflection at 2.96 mrad.

8.3.2 TXRF spectra

Single spectra have been recorded for 100s life time for the 20ng and 500ng Arsenic samples. The exciting energy was tuned to 12200eV. Figure 8.4 shows a comparison of these spectra with spectra recorded for the same samples in GI geometry and at an exciting energy of 12500eV. For the calculation of the detection limits (Limit of Detection, LD) this disagreement in the exciting energies was not corrected as the decrease of the photoelectric cross-section from 12200eV to 12500eV is smaller than 6%. The absorption of the Arsenic fluorescence radiation due to the air between sample and detector (distance 40mm) is smaller than 2% and was therefore also neglected.

Chapter 8   Comparison of Grazing Exit and TXRF geometry for XANES analysis

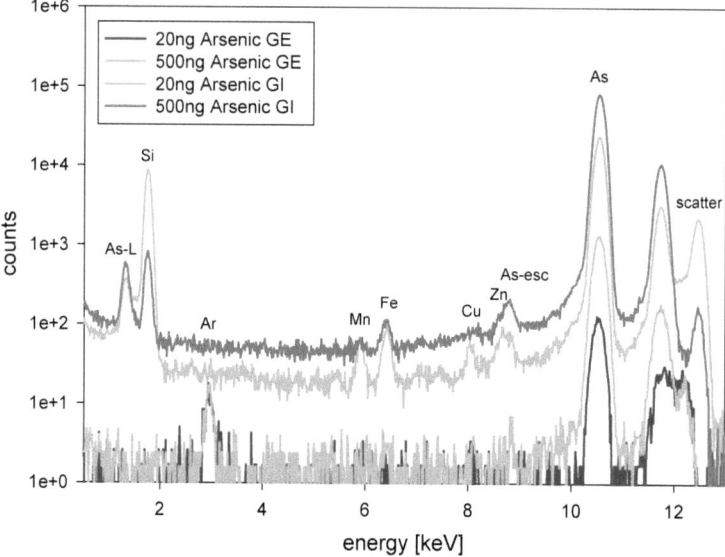

Figure 8.4: Comparison of spectra of identical samples recorded for 100s lifetime in GE and GI geometry respectively. The GE measurements have been performed in vacuum and the GE spectra were recorded in air. Therefore an Argon peak can be seen in the GE spectra while the contribution from the Silicon reflector is missing (absorption in air). The excitation energies were 12200eV and 12500eV for the GE and GI setup respectively.

Detection limits for the 20ng and 500ng Arsenic samples were determined by extrapolation for a 1000 s measuring time and found to be 2 pg and 25 pg for the GI geometry and 55 pg and 200 pg for the GE geometry respectively. This shows that the sensitivity of the GE setup is one order of magnitude lower than the one of the GI geometry. The discrepancy in the detection limits for the 20ng and the 500ng sample can be explained with an increased background for higher sample masses in GI geometry. The trend is confirmed by the LD calculated for the 4ng sample measured in GI geometry which was found to be 650 fg and corresponds to the values determined in chapter 6 (~200fg). Regarding the GE measurements the spectral background is almost zero for both samples. Therefore the difference in the limits of detection is lower (~ a factor 4 instead of a factor 10 for the GI measurements) but still existent. It can not be excluded that both techniques suffer from saturation effects due to large sample masses – this should be topic of further investigations.

8.3.3 Area scans

A series of dried residues with different total amounts of arsenic masses on Quartz reflectors (20ng, 100ng) and a Silicon reflector (500ng) were scanned with the aid of a polycapillary half lens was producing a beam spot of 40µm in diameter. Figure 8.5 shows the results of the measurements in comparison with the data obtained with the confocal microscope (see chapter 7). All fluorescence intensities are given in counts per second (cps), have been corrected for dead time and are normalized to 100mA ring current. The scanning parameters are given in table 8.2. Fluorescence intensities were calculated by region of interest integration because the spectra are practically background free (see figure 8.4). The results show good agreement with the data obtained with the confocal microscope showing the applicability of the GE geometry for spatially resolved analysis.

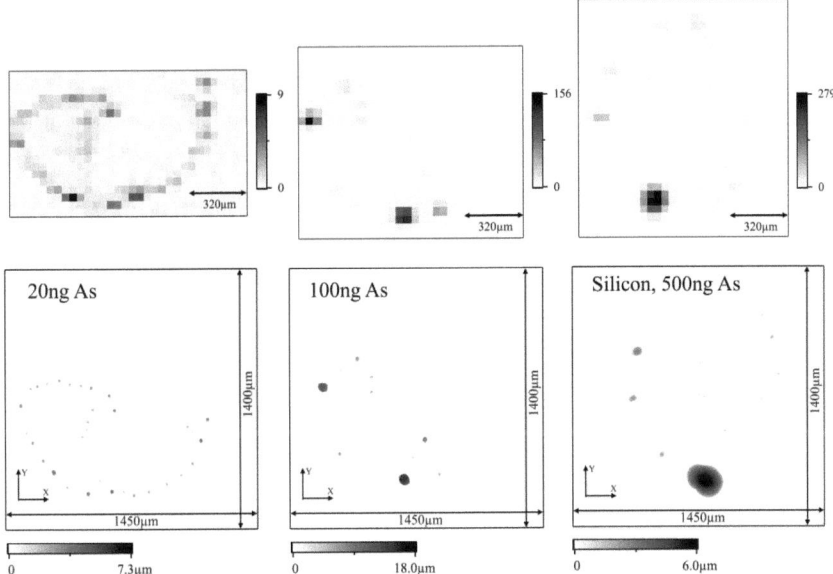

Figure 8.5: Area scans of the 20ng, 100ng and 500ng Arsenic samples (above) in comparison with the results of the measurements with a confocal microscope (below) described in chapter 7. The intensities of the fluorescence maps are given in cps, are dead time corrected and normalized to 100mA ring current.

Scanning parameters	20ng As sample	100ng As sample	500ng As sample
pixels (horizontal x vertical)	32 x 19	30 x 25	27 x 30
Step size	40 µm	40 µm	40 µm
Total scanned area	1280 x 760 µm²	1200 x 1000 µm²	1080 x 1200 µm²

Table 8.2: Parameters of area scans of all measured samples

8.3.4 XANES analysis

Experiments were performed with the aim to study XANES self-absorption effects, which were observed previously for the same samples in GI-XRF geometry. Figure 8.6 shows the results of the XANES measurements performed in GE geometry in comparison with the results of the GI experiments described in section 7.4. The data has been corrected for dead time and was normalized to the flux of the incident radiation with the aid of an ionization chamber. It can be clearly seen that the samples measured using the GE setup show no damping of the oscillations of the absorption coefficient (within the counting statistics). The labels 500ng GE A and 500ng GE B indicate two independent measurements (sets of repetitive scans) which have been performed to check the influence of the counting statistics.

Additionally two samples have been investigated using the focused beam (40µm spot size). Points of maximum fluorescence intensity found during the area scan of the samples have been chosen for XANES analysis. Figure 8.7 indicates the positions in the area maps where the XANES scans have been performed. The results of these measurements in comparison with the experiments performed for the same samples using the unfocused beam and the GI geometry are shown in figure 8.8. A small damping of the white line of the scans recorded using the focused beam can be observed. The effect seems to be stronger for the 500ng sample. Due to the limited time of measurements at a synchrotron only two samples have been investigated using the polycapillary half lens. Therefore a significant conclusion concerning the origin of this effect can not be given. A study of a larger set of samples to see if this damping is inherent would be an interesting topic for further investigations.

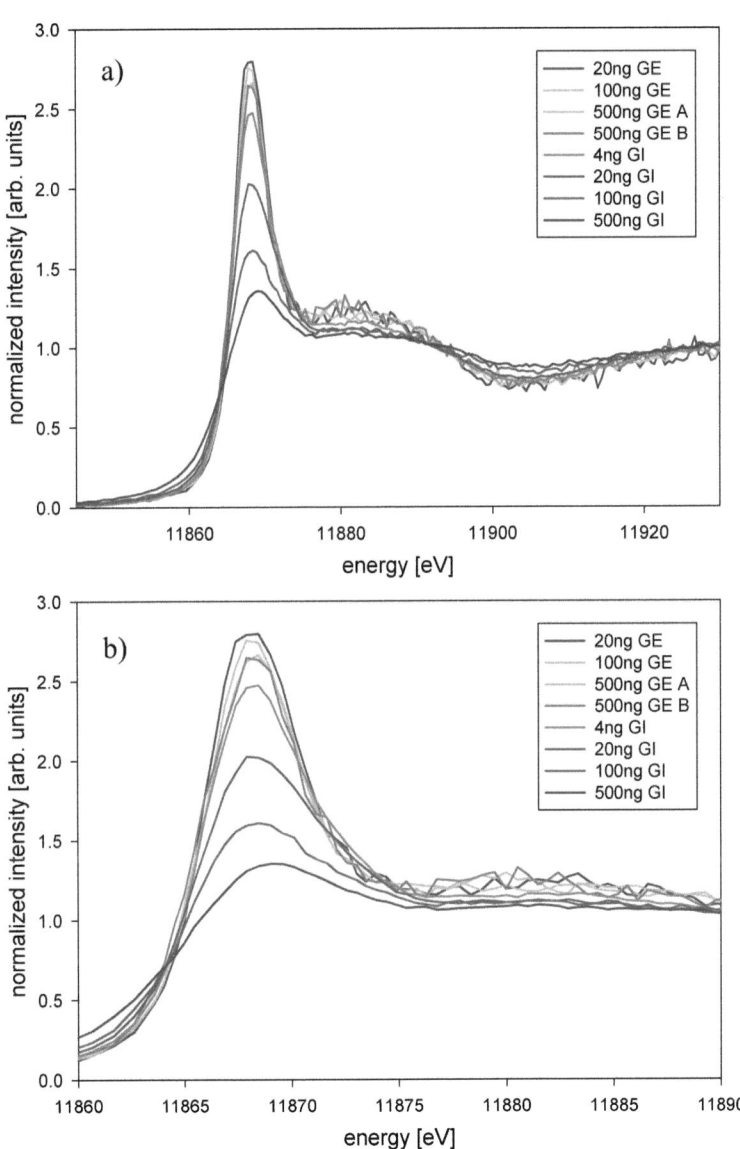

Figure 8.6: Comparison of XANES measurements performed for the same samples in grazing incidence (GI) and grazing exit (GE) geometry. In b) the near edge region is zoomed.

Chapter 8   Comparison of Grazing Exit and TXRF geometry for XANES analysis

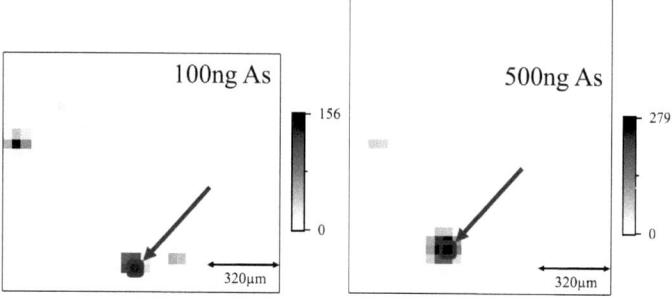

Figure 8.7: Points (blue circles) chosen for XANES analysis of the 100ng and 500ng Arsenic samples using the focused beam.

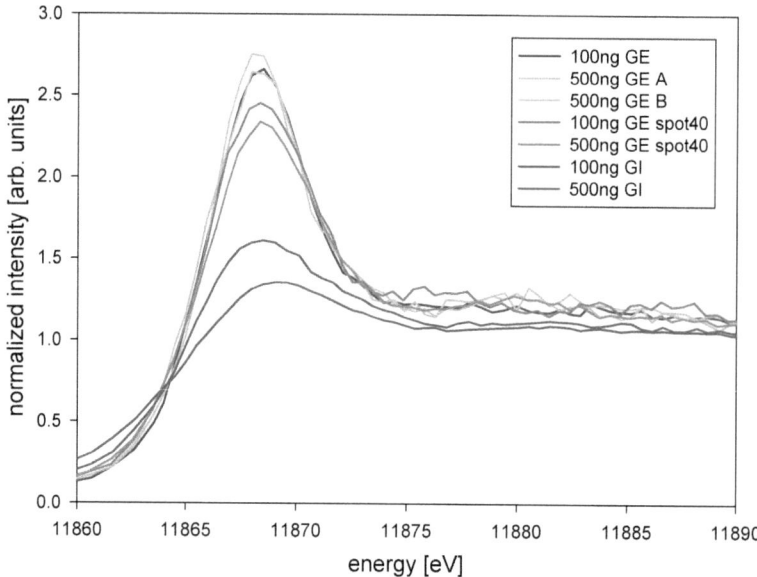

Figure 8.8: XANES region of the 100ng and 500ng Arsenic samples recorded with different setups. The samples labelled with "GI" have been measured using grazing incidence geometry. The rest of the samples were investigated with a grazing exit setup and the curves indicated with "spot40" have been excited by a focused beam with spot size 40μm.

8.4 Summary

It could be shown that the newly developed grazing exit setup utilizing a movable lightweight SDD which allows a fixed sample position works very well for synchrotron radiation induced GE-XRF analysis. Supplementary to the advantages of a TXRF analysis in grazing incidence geometry (multi-element analysis, nondestructive, surface sensitive, determination of contamination type (residual, surface layer, bulk)) the GE experiment allows using a focusing optic for the exciting radiation. This enables spatially resolved investigations of the sample. Angular dependent measurements of bulk and droplet samples as well as area scans of droplet samples with a resolution of 40µm have been performed showing the advantages of the GE setup.

Detection limits calculated to compare the performance of the GE and the GI arrangement showed that the sensitivity of the GE setup is one order of magnitude lower than the one of the GI geometry.

Both geometries can be coupled to X-Ray absorption spectroscopy to gain information on the chemical state of an element of interest. XANES measurements in GE geometry have been performed and compared with results obtained with a GI arrangement. It could be shown that the GE setup does not suffer from self-absorption effects which are typical in the GI experiments. However due to the lower sensitivity it is difficult to apply the GE geometry to XAFS analysis of trace amounts (few ng) of samples. The self-absorption effect in the GI geometry on the other hand decreases rapidly with lower sample amounts. Therefore it would be advantageous for a XAFS analysis to measure higher concentrated standard samples in a GE setup and highly diluted samples in GI geometry.

Chapter 9

Concluding remarks

The previous chapters showed that Total reflection X-Ray Fluorescence (TXRF) analysis in combination with X-ray Absorption Near Edge Structures (XANES) analysis is a powerful and versatile method to perform chemical speciation studies at trace element levels. This combined technique has been successfully applied to various analytical problems. All measurements were performed at HASYLAB beamline L using the synchrotron radiation source DORIS III.

Iron contaminations on Si wafer surface are known to be a serious limiting factor to yield and reliability of complementary metal oxide semiconductor (CMOS) based integrated circuits. Main purpose of the study was to test the method for a contamination issue as it may appear in a microelectronic VLSI (Very-large-scale integration) production fab. It could be shown that SR-TXRF in combination with XAS enables a XANES analysis of wafer surface contaminations even in the pg region within a reasonable time frame. The setup allowed a spatially resolved multi-element analysis of the wafers surface. Additionally the type of the contamination (residual, surface layer, bulk) and the oxidation state of iron in the samples could be determined.

The importance of aerosols not only for human health but also for the cloud formation and the albedo of the earth has become apparent recently. The particle size and the elemental composition of aerosols are two important characteristics to determine their toxicity and provide information about their origin and geo potential. Using SR-TXRF elemental amounts in atmospheric aerosols were determined with respect to the particle size. Due to the excellent features of SR-TXRF detection limits in the range of pg/m^3 could be achieved for 20 minute sampling time. This short time collection allows studying temporal variation of elemental concentrations in size-fractioned aerosols. Evaluation of the size distribution of the elemental concentrations enabled tracing possible particle sources. The detection limits found here will allow for a large-scale study of Pb in the atmosphere and offer the possibility of further speciation e.g. by Pb L-edge XANES. The results of the Fe K-edge SR-TXRF-XANES analysis of the aerosol samples showed that Fe was present in the oxidation state of three (predominately in the form of Iron(III)-oxide) in all collected aerosols. This is in good agreement with other studies on the oxidation state of Fe in aerosols though in rain and cloud

Chapter 9   Concluding remarks

samples high amounts of Fe(II) were reported. The analysis of rain water could be an interesting future topic for SR-TXRF-XANES analysis related to the results of this thesis.

The possibility offered by SR-TXRF-XANES analysis to determine the oxidation state of elements present in very low amounts was also utilized for the investigation of Arsenic in the xylem of cucumber (Cucumis sativus L.) plants. The speciation of arsenic is relevant because the toxicity of arsenic differs considerably dependent on the oxidation state and chemical form and it is known that plants have the capability to change the oxidation state of arsenic. It could be demonstrated that a speciation of As is possible down to the 30 ng/mL level. The results show that cucumber plants treated with arsenate (As(V)) in concentrations of 150 ng/mL convert As(V) to As(III). All results concerning arsenic speciation in xylem saps and nutrient solutions are in good agreement with those obtained by high performance liquid chromatography high resolution inductively coupled plasma mass spectrometry (HPLC-HR-ICP-MS) which indicates the competitive capability of SR-TXRF XANES for trace element speciation.

Owing to the fact that up to now only a few experimental studies have been performed utilizing this combination a major point for all investigations was to show the applicability of the method. One of the challenges which occurred during these studies was the so called "self-absorption effect" due to the extreme grazing incidence geometry of TXRF. This effect results in a damping of the oscillations of the absorption fine structure and was observed when measuring samples with higher concentrations. It was assumed that this effect occurs due to the absorption of the incident beam along its path through the sample (which is especially long for TXRF geometry). The path's length is equivalent to the penetration depth of the incident beam and is therefore energy dependent. This is the reason why this effect appears for (energy dependent) XANES measurements.

Measurements have been compared with theoretical calculations to investigate the influence of self absorption effects on TXRF-XANES analysis. The effect of the sample shape as well as the one of different concentrated droplet samples was studied. The experimental results showed clearly a linear correlation of the damping of the oscillations with the total mass of the samples. A simple Monte-Carlo algorithm was developed to simulate the self absorption effect and the results showed good agreement with the measurements. The presented work proposes a rather simple way to study *a priori* the absorption effects that will show up in TXRF XANES and allow the scientist to prepare the sample according to needs (measuring time, acceptable self absorption) for the actual experiment.

The sample parameters (sample shape) required for the simulations were obtained by measurements with a confocal microscope. The results of these investigations showed large differences for the sample geometry depending on the reflector material and the preparation technique. The shape of dried residues on different reflectors due to different surface characteristics should be topic of further investigations. Another important point related to this topic is the influence of the sample preparation. The deposition of the aqueous sample and the drying process seems to be crucial for quantification in TXRF (especially) without using an internal standard. The best approach to overcome absorption problems seems to be an array of small spots produced with picoliter-pipettes or inkjet printers. This topic should also be further investigated.

Finally the inverse TXRF geometry – the grazing exit (GE) setup – was tested for its applicability to XANES analysis. It was furthermore applied to gain a better understanding of the above mentioned self absorption effect because it was suggested that a normal incidence-grazing-exit geometry would not suffer from self-absorption effects in XAFS analysis due to the minimized path length of the incident beam through the sample. The results proved this assumption and in turn confirmed the occurrence of the self absorption effect for TXRF geometry. However due to its lower sensitivity (one order of magnitude lower than for TXRF) it is difficult to apply the GE geometry to XAFS analysis of trace amounts (few ng) of samples. The self-absorption effect in the TXRF geometry on the other hand decreases rapidly with lower sample amounts. Therefore it could be advantageous for future XAFS analyses to measure higher concentrated standard samples in a GE setup and highly diluted samples in TXRF geometry.

Bibliography

[1] H.G.J. Moseley, The high frequency spectra of the elements, Philosophical Magazine 26 (1913) 1024-1034.

[2] H.G.J. Moseley, The high frequency spectra of the elements Part II, Philosophical Magazine 27 (1914) 703-713.

[3] F.R. Elder, A.M. Gurewitsch, R.V. Langmuir, H.C. Pollock, Radiation from Electrons in a Synchrotron, Physical Review 71 (1947) 829.

[4] H.C. Pollock, The discovery of synchrotron radiation, Am. J. Phys. 51 (1983) 278 - 280.

[5] D. Iwanenko, I. Pomeranchuk, On the Maximal Energy Attainable in a Betatron, Physical Review 65 (1944) 343.

[6] J.P. Blewett, Radiation Losses in the Induction Electron Accelerator, Physical Review 69 (1946) 87.

[7] A. Liénard, Champ électrique et magnétique produit par une charge concentrée en un point et animée d'un mouvement quelconque, L'Éclairage Électrique 16 (1898) 5.

[8] G.A. Schott, Electromagnetic Radiation Cambridge University Press, Cambridge, 1912.

[9] J. Schwinger, Electron Radiation in High Energy Accelerators, Physical Review 70 (1946) 798.

[10] J. Schwinger, On the Classical Radiation of Accelerated Electrons, Physical Review 75 (1949) 1912.

[11] J. Schwinger, A Quantum Legacy: Seminal Papers of Julian Schwinger, in: A.M. Kimball (Ed.), World Scientific, 2000, p. 307.

[12] F.R. Elder, R.V. Langmuir, H.C. Pollock, Radiation from Electrons Accelerated in a Synchrotron, Physical Review 74 (1948) 52.

Bibliography

[13]   D.R. Corson, Radiation by Electrons in Large Orbits, Physical Review 90 (1953) 748.

[14]   I.M. Ado, P.A. Cherenkov, Energy Distribution of Incoherent Radiation from Synchrotron Electrons, Soviet Phys. (Doklady); Vol: 1 (1956) Pages: 517-519.

[15]   D.H. Tomboulian, P.L. Hartman, Spectral and Angular Distribution of Ultraviolet Radiation from the 300-Mev Cornell Synchrotron, Physical Review 102 (1956) 1423.

[16]   E. Koch, Handbook on synchrotron radiation North-Holland Publishing Company, Amsterdam, 1983.

[17]   K. Wille, Synchrotron radiation sources, Reports on Progress in Physics 54 (1991) 1005-1067.

[18]   K. Wille, Physik der Teilchenbeschleuniger und Synchrotronstrahlungsquellen, Teubner Verlag, Stuttgart, 1996.

[19]   J.D. Jackson, Classical electrodynamics 3. ed., Wiley New York, NY [u.a.] 1999.

[20]   H. Wiedemann, Synchrotron Radiation, Springer, Berlin, 2003.

[21]   A. Hofmann, The Physics of Synchrotron Radiation, Cambridge University Press, Cambridge, 2004.

[22]   M. Terasawa, M. Kihara, Basic Characteristics of Synchrotron Radiation and Its Related Facilities and Intrumentation, in: H. Saishi, Y. Gohshi (Eds), Applications of Synchrotron Radiation to Materials Analysis, Elsevier Science B.V., Amsterdam, 1996, pp. 1-78.

[23]   K. Wille, www.nm.physik.uni-dortmund.de/delta/p2000/, 2008.

[24]   W. Gudat, Erzeugung und Eigenschaften von Synchrotronstrahlung, Synchrotronstrahlung in der Festkörperforschung, Vorlesungsmanuskripte des 18.IFF, Kernforschungsanlage Jülich GmbH, Jülich, 1987.

[25]   http://doris.desy.de/.

[26]   E.E. Koch, C. Ranz, E.W. Weiner, Optik 45 (1976) 395.

[27]   K.R. Lea, Highlights of synchrotron radiation, Physics Reports 43 (1978) 337-375.

[28] K. Codling, R.P. Madden, Characteristics of the "Synchrotron Light" from the NBS 180-MeV Machine, Journal of Applied Physics 36 (1965) 380-387.

[29] K. Balewski, W. Brefeld, U. Hahn, J. Pflüger, R. Rossmanith, An Undulator at PETRA II – a New Synchrotron Source at DESY, Proceedings of IEEE PAC '95 (1995) 275.

[30] http://tesla.desy.de, TESLA XFEL - Technical Design Report, DESY Deutsches Elektronen-Synchrotron, Hamburg, 2002.

[31] W. Kossel, Zum Bau der Röntgenspektren, Zeitschrift für Physik 1 (1920) 119-134.

[32] R.d.L. Kronig, Zur Theorie der Feinstruktur in den Röntgenabsorptionsspektren, Zeitschrift für Physik 70 (1931) 317-323.

[33] R. Stumm von Bordwehr A History of X-ray absorption fine structure, Annales de Physique 14 (1989) 377-465.

[34] F. Lytle, The EXAFS family tree: a personal history of the development of extended X-ray absorption fine structure, Journal of Synchrotron Radiation 6 (1999) 123-134.

[35] L.V. Azároff, Theory of Extended Fine Structure of X-Ray Absorption Edges, Reviews of Modern Physics 35 (1963) 1012.

[36] E.A. Stern, Theory of the extended x-ray-absorption fine structure, Physical Review B 10 (1974) 3027.

[37] P.A. Lee, P.H. Citrin, P. Eisenberger, B.M. Kincaid, Extended x-ray absorption fine structure - its strengths and limitations as a structural tool, Reviews of Modern Physics 53 (1981) 769.

[38] E.A. Stern, S.M. Heald, Basic principles and applications of EXAFS, in: E.E. Koch (Ed.), Handbook on Synchrotron Radiation, North-Holland Publishing Company, 1983.

[39] D.C. Koningsberger, R. Prins (Eds.), Xray Absorption: Principles, Applications, Techniques of EXAFS, SEXAFS, and XANES, in Chemical Analysis, John Wiley and Sons, New York, 1988.

[40] J.J. Rehr, R.C. Albers, Theoretical approaches to x-ray absorption fine structure, Reviews of Modern Physics 72 (2000) 621.

Bibliography

[41] D. Norman, X-ray absorption spectroscopy (EXAFS and XANES) at surfaces, Journal of Physics C: Solid State Physics 19 (1986) 3273.

[42] J. Stöhr, NEXAFS Spectroscopy, Springer-Verlag, Berlin, Heidelberg, New York, 1996.

[43] J.J. Rehr, R.C. Albers, C.R. Natoli, E.A. Stern, New high-energy approximation for x-ray-absorption near-edge structure, Physical Review B 34 (1986) 4350.

[44] M. Newville, Fundamentals of XAFS. http://cars9.uchicago.edu/xafs/xas_fun/xas_fundamentals.pdf, 2004.

[45] S.J. Gurman, N. Binsted, I. Ross, A rapid, exact curved-wave theory for EXAFS calculations, Journal of Physics C: Solid State Physics 17 (1984) 143-151.

[46] http://leonardo.phys.washington.edu/feff/, the FEFF Project Home Page (2008).

[47] J.J. Rehr, R.C. Albers, Scattering-matrix formulation of curved-wave multiple-scattering theory: Application to x-ray-absorption fine structure, Physical Review B 41 (1990) 8139.

[48] D.E. Sayers, E.A. Stern, F.W. Lytle, New Technique for Investigating Noncrystalline Structures: Fourier Analysis of the Extended X-Ray-Absorption Fine Structure, Physical Review Letters 27 (1971) 1204.

[49] A.G. McKale, B.W. Veal, A.P. Paulikas, S.K. Chan, G.S. Knapp, Improved ab initio calculations of amplitude and phase functions for extended x-ray absorption fine structure spectroscopy, J. Am. Chem. Soc. 110 (1988) 3763-3768.

[50] http://gnxas.unicam.it, the G N X A S web page (2008).

[51] R.C. Albers, Electronic band-structure methods for XAS, Physica B: Condensed Matter 158 (1989) 372-374.

[52] A.H. Compton, The Total Reflection of X-rays, Philosophical Magazine 45 (1923) 1121-1132.

[53] R.S. Shankland (Ed.), Scientific Papers of Arthur Holly Compton: X-Ray and Other Studies, University of Chicago Press, Chicago, 1973.

[54] Y. Yoneda, T. Horiuchi, Optical Flats for Use in X-Ray Spectrochemical Microanalysis, Review of Scientific Instruments 42 (1971) 1069-1070.

[55] P. Wobrauschek, Totalreflexions Röntgenfluoreszenzanalyse, PhD Thesis (1975), Atominstitut der österreichischen Universitäten, Technical University of Vienna

[56] H. Aiginger, P. Wobrauschek, A method for quantitative X-ray fluorescence analysis in the nanogram region, Nuclear Instruments and Methods 114 (1974) 157-158.

[57] P. Wobrauschek, H. Aiginger, Total-reflection x-ray fluorescence spectrometric determination of elements in nanogram amounts, Anal. Chem. 47 (1975) 852-855.

[58] J. Knoth, H. Schwenke, An X-ray fluorescence spectrometer with totally reflecting sample support for trace analysis at the ppb level, Fresenius' Journal of Analytical Chemistry 291 (1978) 200-204.

[59] H. Schwenke, J. Knoth, A highly sensitive energy-dispersive X-ray spectrometer with multiple total reflection of the exciting beam, Nuclear Instruments and Methods 193 (1982) 239-243.

[60] S.K. Andersen, J.A. Golovchenko, G. Mair, New Applications of X-Ray Standing-Wave Fields to Solid State Physics, Physical Review Letters 37 (1976) 1141.

[61] P.L. Cowan, J.A. Golovchenko, M.F. Robbins, X-Ray Standing Waves at Crystal Surfaces, Physical Review Letters 44 (1980) 1680.

[62] M.J. Bedzyk, G.M. Bommarito, J.S. Schildkraut, X-ray standing waves at a reflecting mirror surface, Physical Review Letters 62 (1989) 1376.

[63] L.G. Parratt, Surface Studies of Solids by Total Reflection of X-Rays, Physical Review 95 (1954) 359.

[64] R.S. Becker, J.A. Golovchenko, J.R. Patel, X-Ray Evanescent-Wave Absorption and Emission, Physical Review Letters 50 (1983) 153.

[65] D.K.G. de Boer, Glancing-incidence x-ray fluorescence of layered materials, Physical Review B 44 (1991) 498.

Bibliography

[66] A. Iida, A. Yoshinaga, K. Sakurai, Y. Gohshi, Synchrotron radiation excited x-ray fluorescence analysis using total reflection of x-rays, Anal. Chem. 58 (1986) 394-397.

[67] P.A. Pella, R.C. Dobbyn, Total reflection energy-dispersive x-ray fluorescence spectrometry using monochromatic synchrotron radiation. Application to selenium in blood serum, Anal. Chem. 60 (1988) 684-687.

[68] S. Brennan, W. Tompkins, N. Takaura, P. Pianetta, S.S. Laderman, A. Fischer-Colbrie, J.B. Kortright, M.C. Madden, D.C. Wherry, Wide band-pass approaches to total-reflection X-ray fluorescence using synchrotron radiation, Nuclear Instruments and Methods in Physics Research Section A: Accelerators, Spectrometers, Detectors and Associated Equipment 347 (1994) 417-421.

[69] F. Hegedüs, P. Wobrauschek, C. Streli, P. Winkler, R. Rieder, W. Ladisich, M. Victoria, R.W. Ryon, W.F. Sommer, Detection of transmutational elements in copper by means of total reflection x-ray fluorescence spectrometry using synchrotron radiation, X-Ray Spectrometry 24 (1995) 253-254.

[70] R. Rieder, P. Wobrauschek, W. Ladisich, C. Streli, H. Aiginger, S. Garbe, G. Gaul, A. Knöchel, F. Lechtenberg, Total reflection X-ray fluorescence analysis with synchrotron radiation monochromatized by multilayer structures, Nuclear Instruments and Methods in Physics Research Section A: Accelerators, Spectrometers, Detectors and Associated Equipment 355 (1995) 648-653.

[71] P. Wobrauschek, R. Gorgl, P. Kregsamer, C. Streli, S. Pahlke, L. Fabry, M. Haller, A. Knochel, M. Radtke, Analysis of Ni on Si-wafer surfaces using synchrotron radiation excited total reflection X-ray fluorescence analysis, Spectrochimica Acta Part B: Atomic Spectroscopy 52 (1997) 901-906.

[72] R. Klockenkämper, Total Reflection X-Ray Fluorescence Analysis, Wiley-Interscience, New York, 1997.

[73] C. Streli, Development of total reflection x-ray fluorescence analysis at the Atominstitute of the Austrian Universities, X-Ray Spectrometry 29 (2000) 203-211.

[74] P. Kregsamer, C. Streli, P. Wobrauschek, Total-Reflection X-ray Fluorescence, in: R. Van Grieken, A. Markowicz (Eds), Handbook of X-Ray Spectrometry, Marcel Dekker, Inc., New York, Basel, 2001, pp. 559 - 602.

Bibliography

[75]   P. Wobrauschek, Total reflection x-ray fluorescence analysis - a review, X-Ray Spectrometry 36 (2007) 289-300.

[76]   F. Meirer, G. Pepponi, C. Streli, P. Wobrauschek, P. Kregsamer, N. Zöger, G. Falkenberg, Parameter study of self-absorption effects in TXRF-XANES analysis of arsenic, submitted to Spectrochimica Acta Part B (2008).

[77]   P. Kregsamer, Röntgenfluoreszenzanalyse von seltenen Erden in Totalreflexionsgeometrie, PhD Thesis (1990), Atominstitut der österreichischen Universitäten, Technical University of Vienna

[78]   R. Rieder, Verbesserung der Nachweisgrenzen bei der Totalreflexions-Röntgenfluoreszenzanalyse durch den Einsatz von Synchrotronstrahlung und Bau einer neuen Meßkammer, PhD Thesis (1994), Atominstitut der österreichischen Universitäten, Technical University of Vienna

[79]   G. Pepponi, Synchrotron Radiation induced Total Reflection X-Ray Fluorescence Analysis applied to Material Science, PhD Thesis (2002), Atominstitut der österreichischen Universitäten, Technical University of Vienna

[80]   B.L. Henke, E.M. Gullikson, J.C. Davis, X-Ray Interactions: Photoabsorption, Scattering, Transmission, and Reflection at E = 50-30,000 eV, Z = 1-92, Atomic Data and Nuclear Data Tables 54 (1993) 181-342.

[81]   H. Saisho, H. Hashimoto, X-Ray Fluorescence Analysis, in: H. Saishi, Y. Gohshi (Eds), Applications of Synchrotron Radiation to Materials Analysis, Elsevier Science B.V., Amsterdam, 1996, pp. 79-169.

[82]   U. Weisbrod, R. Gutschke, J. Knoth, H. Schwenke, X-ray induced fluorescence spectrometry at grazing incidence for quantitative surface and layer analysis Fresenius' Journal of Analytical Chemistry 341 (1991) 83-86.

[83]   D.K.G. De Boer, Angular dependence of X-ray fluorescence intensities, X-Ray Spectrometry 18 (1989) 119-129.

[84]   P. Kregsamer, Fundamentals of total reflection X-ray fluorescence, Spectrochimica Acta Part B: Atomic Spectroscopy 46 (1991) 1332-1340.

[85] R.W. Ryon, J.D. Zahrt, in: R. Van Grieken, A. Markowicz (Eds), Handbook of X-ray Spectrometry, Marcel Dekker, Inc (112001), 1993, pp. 491 - 515.

[86] H. Aiginger, P. Wobrauschek, C. Brauner, Energy-dispersive fluorescence analysis using Bragg-reflected polarized X-rays, Nuclear Instruments and Methods 120 (1974) 541-542.

[87] P.K. de Bokx, H.P. Urbach, Laboratory grazing-emission x-ray fluorescence spectrometer, Review of Scientific Instruments 66 (1995) 15-19.

[88] M. Claes, R. Van Grieken, P. De Bokx, Comparison of Grazing Emission XRF with Total Reflection XRF and Other X-Ray Emission Techniques, X-Ray Spectrometry 26 (1997) 153-158.

[89] R.D. Pérez, H.J. Sánchez, M. Rubio, Theoretical model for the calculation of interference effects in TXRF and GEXRF, X-Ray Spectrometry 30 (2001) 292-295.

[90] K. Tsuji, Z. Spolnik, K. Wagatsuma, S. Nagata, I. Satoh, Grazing-Exit X-Ray Spectrometry for Surface and Thin-Film Analyses, Analytical Sciences 17 (2001) 145-148.

[91] A. Iida, X-ray spectrometric applications of a synchrotron x-ray microbeam, X-Ray Spectrometry 26 (1997) 359-363.

[92] C. Streli, G. Pepponi, P. Wobrauschek, C. Jokubonis, G. Falkenberg, G. Zaray, J. Broekaert, U. Fittschen, B. Peschel, Recent results of synchrotron radiation induced total reflection X-ray fluorescence analysis at HASYLAB, beamline L, Spectrochimica Acta Part B: Atomic Spectroscopy 61 (2006) 1129-1134.

[93] P. Wobrauschek, P. Kregsamer, W. Ladisich, C. Streli, S. Pahlke, L. Fabry, S. Garbe, M. Haller, A. Knochel, M. Radtke, TXRF with synchrotron radiation Analysis of Ni on Si-wafer surfaces, Nuclear Instruments and Methods in Physics Research Section A: Accelerators, Spectrometers, Detectors and Associated Equipment 363 (1995) 619-620.

[94] G. Pepponi, P. Wobrauschek, C. Streli, N. Zöger, F. Hegedüs, Synchrotron radiation-induced TXRF of reactor steel samples, X-Ray Spectrometry 30 (2001) 267-272.

[95] C. Streli, G. Pepponi, P. Wobrauschek, C. Jokubonis, G. Falkenberg, G. Zaray, A new SR-TXRF vacuum chamber for ultra-trace analysis at HASYLAB, Beamline L, X-Ray Spectrometry 34 (2005) 451-455.

[96] P. Pianetta, N. Takaura, S. Brennan, W. Tompkins, S.S. Laderman, A. Fischer-Colbrie, A. Shimazaki, K. Miyazaki, M. Madden, D.C. Wherry, J.B. Kortright, Total reflection x-ray fluorescence spectroscopy using synchrotron radiation for wafer surface trace impurity analysis (invited), Rev. Sci. Instrum. 66 (1995) 1293-1297.

[97] F. Comin, M. Navizet, P. Mangiagalli, G. Apostolo, An industrial SR TXRF facility at ESRF, Nuclear Instruments and Methods in Physics Research Section B: Beam Interactions with Materials and Atoms 150 (1999) 538-542.

[98] P. Pianetta, K. Baur, A. Singh, S. Brennan, J. Kerner, D. Werho, J. Wang, Application of synchrotron radiation to TXRF analysis of metal contamination on silicon wafer surfaces, Thin Solid Films 373 (2000) 222-226.

[99] G. Falkenberg, G. Pepponi, C. Streli, P. Wobrauschek, Comparison of conventional and total reflection excitation geometry for fluorescence X-ray absorption spectroscopy on droplet samples, Spectrochimica Acta Part B: Atomic Spectroscopy 58 (2003) 2239-2244.

[100] J. Kawai, S. Hayakawa, Y. Kitajima, Y. Gohshi, X-ray absorption fine structure (XAFS) of Si wafer measured using total reflection X-rays, Spectrochimica Acta Part B: Atomic Spectroscopy 54 (1999) 215-222.

[101] K. Baur, Brennan S., Pianetta P., Opila R., Looking at Trace Impurities on Silicon Wafers with Synchrotron Radiation, Anal. Chem. 74 (2002) 609A-616A.

[102] A. Singh, Baur K., Brennan S., Homma T., Kubo N. and Pianetta P., X-Ray Absorption Spectroscopy on Copper Trace Impurities on Silicon Wafers, MRS Proceedings 716 (2002).

[103] F. Meirer, G. Pepponi, C. Streli, P. Wobrauschek, V.G. Mihucz, G. Záray, V. Czech, J.A.C. Broekaert, U.E.A. Fittschen, G. Falkenberg, Application of synchrotron-radiation-induced TXRF-XANES for arsenic speciation in cucumber (*Cucumis sativus L.*) xylem sap, X-Ray Spectrometry 36 (2007) 408-412.

[104] G. Pepponi, B. Beckhoff, T. Ehmann, G. Ulm, C. Streli, L. Fabry, S. Pahlke, P. Wobrauschek, Analysis of organic contaminants on Si wafers with TXRF-NEXAFS, Spectrochimica Acta Part B: Atomic Spectroscopy 58 (2003) 2245-2253.

Bibliography

[105] L. Tröger, D. Arvanitis, K. Baberschke, H. Michaelis, U. Grimm, E. Zschech, Full correction of the self-absorption in soft-fluorescence extended x-ray-absorption fine structure, Physical Review B 46 (1992) 3283.

[106] P. Pfalzer, J.P. Urbach, M. Klemm, S. Horn, M.L. denBoer, A.I. Frenkel, J.P. Kirkland, Elimination of self-absorption in fluorescence hard-x-ray absorption spectra, Physical Review B 60 (1999) 9335.

[107] H.J. Sánchez, Direct comparison of total reflection techniques by using a plate beamguide, X-Ray Spectrometry 31 (2002) 145-149.

[108] A.A. Istratov, H. Hieslmair, E.R. Weber, Iron contamination in silicon technology, Applied Physics A: Materials Science & Processing 70 (2000) 489-534.

[109] S. Pahlke, L. Fabry, L. Kotz, C. Mantler, T. Ehmann, Determination of ultra trace contaminants on silicon wafer surfaces using total-reflection X-ray fluorescence TXRF 'state-of-the-art', Spectrochimica Acta Part B: Atomic Spectroscopy 56 (2001) 2261-2274.

[110] www.technos-intl.com/trex630.php, The Technos international Homepage, 2008.

[111] www.rigaku.com/semi/txrf-v300.html, The Rigaku homepage, 2008.

[112] L. Fabry, S. Pahlke, L. Kotz, G. Tölg, Trace-analytical methods for monitoring contaminations in semiconductor-grade Si manufacturing, Fresenius' Journal of Analytical Chemistry 349 (1994) 260-271.

[113] G. Settembre, E. Debrah, Using VPD ICP-MS to monitor trace metals on unpatterned wafer surfaces, Micro 16 (1998) 79.

[114] www.radiantdetectors.com/vortex.html, The SII Nanotechnology Incorporated Website, 2007.

[115] G. Falkenberg, Characterization of a Radiant Vortex Silicon Multi-Cathode X-ray Spectrometer for (total reflection) X-ray fluorescence applications, Hasylab Internal Report 2004 (2005).

[116] www.iaea.or.at/programmes/ripc/physics/faznic/qxas.htm.

[117] http://cars9.uchicago.edu/ifeffit/, The IFEFFIT homepage. 2007.

[118] M. Newville, IFEFFIT: interactive XAFS analysis and FEFF fitting, Journal of Synchrotron Radiation 8 (2001) 322-324.

[119] B. Ravel, M. Newville, ATHENA, ARTEMIS, HEPHAESTUS: data analysis for X-ray absorption spectroscopy using IFEFFIT, Journal of Synchrotron Radiation 12 (2005) 537-541.

[120] Y. Mori, K. Uemura, Y. Iizuka, Whole-Surface Analysis of Semiconductor Wafers by Accumulating Short-Time Mapping Data of Total-Reflection X-ray Fluorescence Spectrometry, Anal. Chem. 74 (2002) 1104-1110.

[121] Y. Mori, K. Uemura, H. Kohno, M. Yamagami, T. Yamada, K. Shimizu, Y. Onizuka, Y. Iizuka, Detection of unknown localized contamination on silicon wafer surface by sweeping-total reflection X-ray fluorescence analysis, Spectrochimica Acta Part B: Atomic Spectroscopy 59 (2004) 1277-1282.

[122] H. Schwenke, J. Knoth, Handbook of X-ray Spectrometry, Marcel Dekker, 1993.

[123] J. Osan, S. Török, V. Groma, C. Streli, P. Wobrauschek, F. Meirer, G. Falkenberg, Trace element analysis of fine aerosol particles with high time resolution using SR-TXRF, HASYLAB Annual Report (2005).

[124] V. Groma, F. Meirer, J. Osan, S. Török, C. Streli, P. Wobrauschek, G. Falkenberg, Trace element analysis of urban and background aerosols using SR-TXRF, HASYLAB Annual Report (2006).

[125] V. Groma, J. Osan, A. Alsecz, S. Török, F. Meirer, C. Streli, P. Wobrauschek, G. Falkenberg, Trace element analysis of airport related aerosols using SR-TXRF, IDÕJÁRÁS - Journal of the Hungarian Meteorological Service 112, accepted (2008).

[126] U.E.A. Fittschen, F. Meirer, C. Streli, P. Wobrauschek, G. Falkenberg, G. Pepponi, J.A.C. Broekaert, J. Thiele, Characterization of Atmospheric Aerosols using SR-TXRF and Fe K-edge TXRF-XANES, HASYLAB Annual Report (2007).

[127] U.E.A. Fittschen, F. Meirer, C. Streli, P. Wobrauschek, J. Thiele, G. Falkenberg, G. Pepponi, Characterization of Atmospheric Aerosols using SR-TXRF and Fe K-edge TXRF-XANES, submitted to Spectrochimica Acta Part B (2008).

[128] V. Groma, B. Alföldy, J. Osán, S. Kugler, M. Kalocsai, Impact of the airport related traffic on the urban particulate pollution, Proceeding of European Aerosol Conference in Salzburg, Austria (9th-14th Sept. 2007).

[129] M. Kampa, E. Castanas, Human health effects of air pollution, Environmental Pollution 151 (2008) 362-367.

[130] K. Katsouyanni, G. Touloumi, E. Samoli, A. Gryparis, A. Le Tertre, Y. Monopolis, G. Rossi, D. Zmirou, F. Ballester, A. Boumghar, H.R. Anderson, B. Wojtyniak, A. Paldy, R. Braunstein, J. Pekkanen, C. Schindler, J. Schwartz, Confounding and Effect Modification in the Short-Term Effects of Ambient Particles on Total Mortality: Results from 29 European Cities within the APHEA2 Project, Epidemiology 12(5) (2001) 521-531.

[131] J.E. Penner, M. Andreae, H. Annegarn, L. Barrie, J. Feichter, D. Hegg, A. Jayaraman, R. Teaitch, D. Murphy, J. Nganga, G. Pitari, Aerosols, their direct and indirect effects, in: J.T. Houghton, e. al. (Eds), Climate Change 2001 – the scientific basis, Contribution of Working Group I to the Third Assessment Report of the Intergovernmental Panel on Climate Change, Cambridge University Press, Cambridge, UK, 2001, pp. 289-348.

[132] J. Jaworski, Lead, in: T.C. Hutchinson, K.M. Meema (Eds), SCOPE 31: Lead, Mercury, Cadmium and Arsenic in the Environment, John Wiley & Sons, Chichester, New York, Brisbane, Toronto, 1987, pp. 3-16.

[133] R.M. Harrison, W.T. Sturges, The measurement and interpretation of Br/Pb ratios in airborne particles, Atmospheric Environment (1967) 17 (1983) 311-328.

[134] L.J. Spokes, T.D. Jickells, B. Lim, Solubilisation of aerosol trace metals by cloud processing: A laboratory study, Geochimica et Cosmochimica Acta 58 (1994) 3281-3287.

[135] Y. Erel, S.O. Pehkonen, H. Hoffmann, Redox chemistry of iron in fog and stratus clouds, Journal of Geophysical Research 98 (1987) 18423–18434.

[136] M.O. Andreae, P.J. Crutzen, Atmospheric Aerosols: Biogeochemical Sources and Role in Atmospheric Chemistry, Science 276 (1997) 1052-1058.

[137] K.R. May, An "ultimate" cascade impactor for aerosol assessment, Journal of Aerosol Science 6 (1975) 413-416.

[138] U.E.A. Fittschen, S. Hauschild, M.A. Amberger, G. Lammel, C. Streli, S. Förster, P. Wobrauschek, C. Jokubonis, G. Pepponi, G. Falkenberg, J.A.C. Broekaert, A new technique for the deposition of standard solutions in total reflection X-ray fluorescence spectrometry (TXRF) using pico-droplets generated by inkjet printers and its applicability for aerosol analysis with SR-TXRF, Spectrochimica Acta Part B: Atomic Spectroscopy 61 (2006) 1098-1104.

[139] F. Meirer, C. Streli, P. Wobrauschek, N. Zoeger, C. Jokubonis, G. Pepponi, G. Falkenberg, J. Osan, S. Török, V. Groma, U.E.A. Fittschen, J.A.C. Broekaert, V.G. Mihucz, G. Zaray, V. Czech, J. Hofstätter, P. Roschger, R. Simon, Recent results of the ATI X-Ray group with SR-XRF, XRF Newsletter of the IAEA Laboratories 12 (2006) 7-13.

[140] D.K.G. de Boer, Fundamental parameters for X-ray fluorescence analysis, Spectrochimica Acta Part B: Atomic Spectroscopy 44 (1989) 1171-1190.

[141] V.-M. Kerminen, R. Hillamo, K. Teinilä, T. Pakkanen, I. Allegrini, R. Sparapani, Ion balances of size-resolved tropospheric aerosol samples: implications for the acidity and atmospheric processing of aerosols, Atmospheric Environment 35 (2001) 5255-5265.

[142] E. Mészáros, T. Barcza, A. Gelencsér, J. Hlavay, G. Kiss, Z. Krivácsy, A. Molnár, K. Polyák, Size distributions of inorganic and organic species in the atmospheric aerosol in Hungary, Journal of Aerosol Science 28 (1997) 1163-1175.

[143] J. Osán, B. Alföldy, S. Török, R. Van Grieken, Characterisation of wood combustion particles using electron probe microanalysis, Atmospheric Environment 36 (2002) 2207-2214.

[144] I. Tesseraux, Risk factors of jet fuel combustion products, Toxicology Letters 149 (2004) 295-300.

[145] N. Bukowiecki, M. Hill, R. Gehrig, C.N. Zwicky, P. Lienemann, F. Hegedus, G. Falkenberg, E. Weingartner, U. Baltensperger, Trace Metals in Ambient Air: Hourly Size-Segregated Mass Concentrations Determined by Synchrotron-XRF, Environ. Sci. Technol. 39 (2005) 5754-5762.

[146] G. Lammel, A. Röhrl, H. Schreiber, Atmospheric Lead and Bromine in Germany - Post-abatement Levels, Variabilities and Trends, Environmental Science and Pollution Research 9 (2002) 397-404.

Bibliography

[147] P. Hoffmann, A.N. Dedik, J. Ensling, S. Weinbruch, S. Weber, T. Sinner, P. Gütlich, H.M. Ortner, Speciation of iron in atmospheric aerosol samples, Journal of Aerosol Science 27 (1996) 325-337.

[148] M. Hirabayashi, M. Matsuo, Characterization of Iron in Airborne Particulate Matter by X-ray Absorption Fine Structure Technique, Analytical Sciences 17 (2001) i1581-i1584.

[149] G. Zhang, Atmospheric particulate matter studied by Mössbauer spectroscopy and XAFS, Hyperfine Interactions 151-152 (2003) 299-306.

[150] G. Zhuang, Z. Yi, G.T. Wallace, Iron(II) in rainwater, snow, and surface seawater from a coastal environment, Marine Chemistry 50 (1995) 41-50.

[151] R.J. Kieber, B. Peake, J.D. Willey, B. Jacobs, Iron speciation and hydrogen peroxide concentrations in New Zealand rainwater, Atmospheric Environment 35 (2001) 6041-6048.

[152] I. Varsányi, Z. Fodré, A. Bartha, Arsenic in drinking water and mortality in the Southern Great Plain, Hungary, Environmental Geochemistry and Health 13 (1991) 14-22.

[153] V. Czech, V.G. Mihucz, P. Czövek, E. Cseh, G. Záray, Investigation of arsenate phytotoxicity in cucumber plants, Metal Ions in Biology and Medicine 9 (2006) 158-162.

[154] V. Mihucz, E. Tatár, I. Virág, E. Cseh, F. Fodor, G. Záray, Arsenic speciation in xylem sap of cucumber (*Cucumis sativus* L.), Analytical and Bioanalytical Chemistry 383 (2005) 461-466.

[155] World-Health-Organization, Guidelines for Drinking-water Quality, Guidelines for Drinking-water Quality Vol. 1, 3rd edition (2004).

[156] M. Bissen, Fritz H. Frimmel, Arsenic - a Review. Part I: Occurrence, Toxicity, Speciation, Mobility, Acta hydrochimica et hydrobiologica 31 (2003) 9-18.

[157] R. Chen, B.W. Smith, J.D. Winefordner, M.S. Tu, G. Kertulis, L.Q. Ma, Arsenic speciation in Chinese brake fern by ion-pair high-performance liquid chromatography-inductively coupled plasma mass spectroscopy, Analytica Chimica Acta 504 (2004) 199-207.

[158] A.A. Meharg, J. Hartley-Whitaker, Arsenic uptake and metabolism in arsenic resistant and nonresistant plant species, New Phytologist 154 (2002) 29-43.

[159] C.I. Ullrich-Eberius, A. Sanz, A.J. Novacky, Evaluation of Arsenate- and Vanadate-Associated Changes of Electrical Membrane Potential and Phosphate Transport in Lemna gibba G1, J. Exp. Bot. 40 (1989) 119-128.

[160] A.A. Meharg, M.R. Macnair, Suppression of the High Affinity Phosphate Uptake System: A Mechanism of Arsenate Tolerance in Holcus lanatus L, J. Exp. Bot. 43 (1992) 519-524.

[161] G.M. Kertulis, L.Q. Ma, G.E. MacDonald, R. Chen, J.D. Winefordner, Y. Cai, Arsenic speciation and transport in Pteris vittata L. and the effects on phosphorus in the xylem sap, Environmental and Experimental Botany 54 (2005) 239-247.

[162] T. Gasparics, V.G. Mihucz, E. Tatar, G. Zaray, Hyphenated technique for investigation of nickel complexation by citric acid in xylem sap of cucumber plants, Microchemical Journal 73 (2002) 89-98.

[163] S.M. Webb, J.F. Gaillard, L.Q. Ma, C. Tu, XAS Speciation of Arsenic in a Hyper-Accumulating Fern, Environ. Sci. Technol. 37 (2003) 754-760.

[164] P.G. Smith, I. Koch, R.A. Gordon, D.F. Mandoli, B.D. Chapman, K.J. Reimer, X-ray Absorption Near-Edge Structure Analysis of Arsenic Species for Application to Biological Environmental Samples, Environ. Sci. Technol. 39 (2005) 248-254.

[165] H. Castillo-Michel, J.G. Parsons, J.R. Peralta-Videa, A. Martinez-Martinez, K.M. Dokken, J.L. Gardea-Torresdey, Use of X-ray absorption spectroscopy and biochemical techniques to characterize arsenic uptake and reduction in pea (Pisum sativum) plants, Plant Physiology and Biochemistry 45 (2007) 457-463.

[166] J. Osán, B. Török, S. Török, K.W. Jones, Study of Chemical State of Toxic Metals During the Life Cycle of Fly Ash Using X-Ray Absorption Near-Edge Structure, X-Ray Spectrometry 26 (1997) 37-44.

[167] F. Goodarzi, F.E. Huggins, Speciation of Arsenic in Feed Coals and Their Ash Byproducts from Canadian Power Plants Burning Sub-bituminous and Bituminous Coals, Energy Fuels 19 (2005) 905-915.

[168] http://www.nanofocus.info, The Nanofocus AG homepage, 2007.

[169] www.imagemet.com, The Image Metrology A/S homepage, 2007.

[170] H. Ebel, R. Svagera, M.F. Ebel, A. Shaltout, J.H. Hubbell, Numerical description of photoelectric absorption coefficients for fundamental parameter programs, X-Ray Spectrometry 32 (2003) 442-451.

[171] M. Makoto, K. Yoshiharu, Twisted GFSR generators II, ACM Trans. Model. Comput. Simul. 4 (1994) 254-266.

[172] R.D. Deegan, O. Bakajin, T.F. Dupont, G. Huber, S.R. Nagel, T.A. Witten, Capillary flow as the cause of ring stains from dried liquid drops, Nature 389 (1997) 827-829.

[173] R.D. Deegan, Pattern formation in drying drops, Physical Review E 61 (2000) 475.

[174] R.D. Deegan, O. Bakajin, T.F. Dupont, G. Huber, S.R. Nagel, T.A. Witten, Contact line deposits in an evaporating drop, Physical Review E 62 (2000) 756.

[175] Y.O. Popov, Evaporative deposition patterns: Spatial dimensions of the deposit, Physical Review E (Statistical, Nonlinear, and Soft Matter Physics) 71 (2005) 036313-036317.

[176] T. Noma, H. Miyata, S. Ino, Grazing Exit X-Ray Fluorescence Spectroscopy for Thin-Film Analysis Jpn. J. Appl. Phys. 31 (1992) L900 - L903.

[177] T. Noma, A. Iida, K. Sakurai, Fluorescent-x-ray-interference effect in layered materials, Physical Review B 48 (1993) 17524.

[178] T. Noma, A. Iida, Surface analysis of layered thin films using a synchrotron x-ray microbeam combined with a grazing-exit condition, Review of Scientific Instruments 65 (1994) 837-844.

[179] K. Tsuji, K. Wagatsuma, T. Oku, Glancing-incidence and glancing-takeoff x-ray fluorescence analysis of Ni-GaAs interface reactions, X-Ray Spectrometry 29 (2000) 155-160.

[180] K. Tsuji, S. Sato, K. Hirokawa, Surface-sensitive x-ray fluorescence analysis at glancing incident and takeoff angles, Journal of Applied Physics 76 (1994) 7860-7863.

VDM Verlagsservicegesellschaft mbH

Die VDM Verlagsservicegesellschaft sucht für wissenschaftliche Verlage abgeschlossene und herausragende

Dissertationen, Habilitationen, Diplomarbeiten, Master Theses, Magisterarbeiten usw.

für die kostenlose Publikation als Fachbuch.

Sie verfügen über eine Arbeit, die hohen inhaltlichen und formalen Ansprüchen genügt, und haben Interesse an einer honorarvergüteten Publikation?

Dann senden Sie bitte erste Informationen über sich und Ihre Arbeit per Email an *info@vdm-vsg.de*.

Sie erhalten kurzfristig unser Feedback!

VDM Verlagsservicegesellschaft mbH
Dudweiler Landstr. 99   Telefon +49 681 3720 174
D - 66123 Saarbrücken   Fax   +49 681 3720 1749
www.vdm-vsg.de

Die VDM Verlagsservicegesellschaft mbH vertritt

Printed by Books on Demand GmbH, Norderstedt / Germany